HISTORIC
RHODE ISLAND FARMS

ROBERT A. GEAKE

THE History PRESS

Published by The History Press
Charleston, SC 29403
www.historypress.net

Cover images courtesy of the author.

First published 2013

Manufactured in the United States

ISBN 978.1.62619.262.1

Library of Congress CIP data applied for.

This volume is dedicated to the memory of Harriet W. Kenney, my aunt and an avid amateur photographer of rural Rhode Island. A few of her pictures grace this book, and I attribute my own passion for roadside photography entirely to her influence. Her Rambler Nash was pulled off road on many occasions so that she could take a photograph of an old house, a mill or a farm during the 1950s and early 1960s. Some of the places in the photographs are recognizable to me. Many are gone, though I often found myself looking for them on my travels as I wrote this humble history of Rhode Island's farms.

Contents

Acknowledgements

First and foremost I must thank Jeff Saraceno, former editor at the History Press, who first proposed this book to me. His initial guidance and the editorial work of Katie Orlando made the experience of researching and writing this book a pleasure.

I am, as always, indebted to those historians and historical preservationists who have previously researched and detailed the history of historic farms and properties in the towns and villages of the state of Rhode Island. The reader will find that those studies published by the Rhode Island Historic Preservation Commission from the 1970s into the 1990s proved to be invaluable resources of farm history and stark reminders as I drove around in search of these farms of how much has been lost in the proceeding decades, although I was surprised at how much does remain.

Other information on early farms was gleaned from the memoirs and histories of prominent early Rhode Islanders, as well as those local town histories written by Rhode Island authors. I hope readers will refer to the bibliography to learn more.

I also want to thank the farmers who consented to being interviewed during the course of this book. As outspoken as their nature tends to be on preservation and land use issues within their communities, they are a private people when it comes to their own lives and accomplishments.

Carroll and Molly Harrington were extremely gracious in answering questions about their own history on the farm, and their neighbors, the Dahlquists, were also accommodating in providing a history of their farm, including a well-written report by a daughter for her high school history class in the 1970s.

Historic New England farm managers Don and Heather Minto, as well as their border collies, were extremely social hosts during my visit on a sweltering day, giving me a tour of the farm and great conversation over a glass of cool lemonade at the farmhouse table. James Dame gave a reluctant, but ultimately heartfelt testimony of the fate of his family farm.

Many others were simply tolerant of my sudden request to take pictures of their barns and houses and my hurried note-taking as I learned a little history of their farm—notes that I would later use to find any written records of the property or stories associated with the place.

Dan Santos of Historic New England and the staff and volunteers of Watson Farm, Casey Farm and Eleazer Arnold House were of great help in further understanding the evolution of these historic sites and also their present use as preserved farmlands.

The staff and volunteers of the Newport Preservation society at the Prescott Farm were equally helpful, as were the volunteers at Coggeshall Farm and Smith's Castle during my visits.

My friends at the Warwick Historical Society provided a valuable cache of pictures and newspaper articles, as well as hand-typed accounts of local farms. Warwick's historian laureate Henry A.L. Brown was, as always, a treasure-trove of information and reminiscences of the many farms that once lay on the landscape.

I want also to thank my brother Bill, and friends James Allen and Richard Rubinstein for their assistance in traveling to and recording some of those historic farms, barns and outbuildings that remain in Rhode Island.

And last but not least, I want to mention that as with other New England communities, Rhode Island abounded with farms during its formative years. Any omission of a historic farm is my error, but I hope that in the examples given, I am able to portray a general history and a glimpse of a life shared by so many for so long in our state.

1
Early Farms and Husbandry

The Evolution of the Barnyard

There were few barns in those early years of husbandry in the colony of Rhode Island. Informal lots were drawn for land, and the settlers of Providence—the followers of minister Roger Williams—built houses and a few outbuildings across the lane from the river, where some also built warehouses and docks.[1] They herded their sheep across the narrow crossing that had been used by Native Americans for generations to a place they named Weybosset meadows. This area is also where the settlers planted the crops to sustain the settlement, and with hard work, they harvested enough to sell at market as well. By their second season, the inhabitants were shipping corn and swine to Boston and Salem on the sloops of John Gardner and John Throckmorton.

While sheep were confined to the "medows," hogs roamed freely through the settlement, causing much damage to the crops, as well as becoming a nuisance on the riverbanks by rooting and foraging for the shellfish that some relied on for trade or sustenance. While hogs, cattle and even horses could forage in the common meadows and woods quite well for themselves, as their numbers increased, so did complaints and often violent confrontations.

Sometime before October 1640, John Field petitioned the town assembly to assert that the problem of unpenned livestock was causing great "contention, disunion, and unquietness" in Providence. On at least one occasion, a townsman angered at the destruction of his crops had gone after the offending animals' owner with a club. Mr. Field also admonished his fellow settlers who were "Brawling Constantly in mr. william's medow."[2]

The Sayles House, pictured here in a nineteenth-century photograph, shows the evolution of the early farmhouse. *Courtesy of the Rhode Island Historical Society.*

The first settlers were bound by the law they created as landowners to "improve their grounds at present granted to them viz by preparing to fence[,] to plant[,] to build[,] etc." They could not sell their lot to "any person but an inhabitant" without the consent of the other townsmen, and though timber was plentiful, the assembly decreed that two men should be appointed to "view the timber on ye common that such as have occasion to use timber should report unto them for their advise and consent to fell timber for their use between the shares graunted and mile end cove."[3] Thus, the founders of Providence were apparently wary of outsiders and worried that their own supply of timber would be depleted. It's little wonder that a ruined field often led to fisticuffs.

Thomas Clemence, a friend of Williams, established his small farm on the western boundary of the lands Williams had been deeded from the Narragansett. Clemence purchased the land himself from Winnesaqua, a local sachem who likely had oversight of the nearby stone foundries on Neutonkonecut Hill. His son Richard would build the simple four-room, stone-end farmhouse that still stands as an example of the homesteads established in those first years of the colony. Clemence found, as had William Blackstone, that fewer neighbors made up for poor fences, and thus, he established his own peaceable kingdom within the wide boundaries of early Providence.

With the acquisition of Aquidneck Island in 1638 and the founding of Pocasset, husbandry began to improve. Original founder John Clarke would write that the settlement of Aquidneck came about in a circuitous route, one that began with himself and the others leaving Boston to head north and then south, sailing past "that long and dangerous Cape" and finding Providence. Some still eyed Long Island or the long-distance shores of Delaware for a settlement, but Roger Williams, by whom they "were courteously and lovingly received," advised the company of "two places before us in the same Narraganset Bay, the one upon the Main called Sowames, the other called Acquedneck, now Rode Island."[4]

The garden and grain rights for Sowames was held by Plymouth, a debt the group was unwilling to bear from the outset, and so with Williams's advisement, appealed to authorities in Plymouth for the grant of the island. When this was accomplished, Clarke wrote:

The Clemence-Irons house, pictured here, is an example of an early Rhode Island farmhouse. *Courtesy of Historic New England.*

> *We were now on the wing, and were resolved through the help of Christ, to be clear of all, and be of ourselves.*

In one of the settlements' first town meetings, an ordinance passed that "every one of this body shall have for its present use one acre of meadow for a beast, one acre for five sheep, and one acre and a half for a horse." This act was later repealed, and founders were given more land and the opportunity to own more livestock. Pocasset may have striven to retain civic order through more orderly husbandry than Providence had shown, but on occasion, similar scenes occurred like those that played out in Mr. Williams's meadow; as in the case from September 1638 in which

> *by virtue of a Warrant, George Willmore, George Parker, John Lutner, John Arnold, Samuell Smith, Robert Stanton, Anthony Robinson, John Vahun, being summoned to appeare before the Body for a Riott of drunkenesse by them committed on the 13th…It is further ordered, that Mr. Esson Mr. Coggeshall, & Mr. Willbore shall view such damages that are done upon the Corn & other fruits & accordingly shall give information to the Body.*

Nicholas Esson would later acquire from John Porter and John Samfford, under authority of the town, "sufficient accommodations for four Cowes & planting ground as they shall think meett all wch is for the setting up of a Water Millur wch the sd Mr. Esson hath undertaken to build for the necessary use & good of the plantation."

As the population of men and livestock continued to grow, the town took pains to deal with the issues that naturally arise when both are crowded together. The town leaders ordered that "the swinn that are on the Island shall be sent away from the Plantation six miles up into the island or unto some islands adjacente by the 10th of the 2nd 1639 or else be shutt up that so they may be inoffensive to the towne."

The following spring, as the hogs had been moved away, Mr. Esson, Coggeshall and Willbore were again called on to "take a view of the several damages by the cattle of severall heards of cattle" on the island. The men were to find the owners of the offending animals and report their names, with witnesses to the body. Once found guilty by Judge Coddington and the assembled body, a heavy fine would be imposed on the farmer, who would be jailed if he refused to pay.

Finally in May 1640, the town, now called Portsmouth, ordered that "a place for the impounding of cattle shall be made & sett up in a convenient place of each towne within 30 days."

The other town referred to in the records was Newport. The area that became the town was first settled in May 1639 by Nicholas Easton, who built a house there on the path that would later be called Marborough Street, within sight of the harbor. He was followed soon after by William Coddington, who, having set the affairs of Pocasset in order from its beginning, had bristled under the tongue-lashing of Anne Hutchinson and her followers and had been unceremoniously ousted from his judgeship on the body.

Coddington sailed from the town on his own vessel, and within seven months with the aid of Roger Williams, he had procured the grazing rights to Block Island and asserted his influence in the new assembly, which ordained in December 1639 that "sufficient fences viz [either] hedge or post & raile made about the corn grounds that shal be planted or saune by the 1st of May next & if any man shall be found delinquent therin he shall forfeit for every twenty rod that is defective the Sum of 3 s. 4 d." Most importantly perhaps, for the civil peace desired, the council ordered that "no man" of Newport "shall keep any Hoggs about the Towne exept it be within his own enclosure after the 15th of April until the 15th of October."[5]

Carl Bridenbaugh, the foremost chronicler of colonial life in southeastern New England, wrote of the importance of these islands in the developing agriculture of the colony: "Many of these islands had some open fields or meadows, in which grass grew luxuriantly…much of the best farmland was located on the islands or along the western shores of the Bay, which in places had been cleared about eight miles inland, for as soon as the natives exhausted the soil they moved away, and grass grew in the abandoned fields…the principle advantage of the islands, however, was that they were a safe land on which to raise all kinds of livestock."[6]

In 1645, the English agriculturist Dr. Robert Child sent a description of "Rhode Isle" to his friend Samuel Hartlib, a well-known writer on farming in his country, which read in part, "This place abounds with corn and cattle, especially sheep, there being nigh 1,000 on the Isle."[7]

As early as 1649, Rhode Island was shipping cattle to Boston, as well as to distant Barbados. As more land was cleared and acres were sown with English seed, larger herds were put out to graze. As the cattle increased, Coddington, William Brenton and Mr. Brinley "divided their herds and each put half on the Islands to live in the open." When Block Island became part of the colony in 1672, it also became an active competitor in the sheep and cattle trade.

William Brenton had acquired nearly 2,000 acres by the mid-1660s at a place that became known as Brenton's Neck, a large swath of land that extended from the present area of Fort Adams in the northwest to Bailey's Beach in the southeast. He also owned a 256-acre tract on Conanicut Island and became known as the "leading grazier of New England," amassing 1,600 sheep by the time of his death in 1674.

His son Jaleel inherited Hammersmith Farms and renamed the eastern portion Rocky Farms, building a large stone barn to house the many sheep his father had left him. He later built a house nearby for the tenant herder.

Sheep were herded and shipped on deck with the cattle from these early farms and also supplied the much-needed wool to provide clothing for families. Horses were also brought into the colony early on, though we'll explore their development there later. The invaluable animal on any working farm was the ox.

In the colonial period, these were the animals of choice for plowing the rocky soil, and the "lumbering beasts" were often used for hauling, on a flat sled, stone or lumber for use in building walls or foundations. Oxen also hauled the heavier wagonloads on the farm.

While oxen were essential in working the farm, "neat" cattle brought in the revenue needed to keep the farm running. Cattle in Rhode Island was initially brought over from neighboring Massachusetts in herds of varying English stock. The climate of Narragansett proved to be well suited for raising cattle, and within a few years, by breeding on these larger farms, a native strain began to emerge. These became prized in England, where they "do well unsheltered whole winters."[8]

The English custom of husbandry often left animals in the open, and barns, as they were built, were used exclusively for storing grains. What shelter was given to sheep or cattle were likely "sheep folds" or lean-tos built off the least exposed wall of an enclosure.

In the New England colonies, farmers were finding that adaptations were needed with the shorter growing season and the necessity of sheltering their livestock during the more severe winters. Fortunately, the islands and mainland had "noble stands of timber" that were used at once for the expanding farms outside Providence. Bridenbaugh notes:

> *On Rhode Island from 1638–1660 and again from 1676–1680, most of* [the timber] *went into the construction of houses, barns, and other*

> *farm structures. Barns came second only to dwellings and were often built on the largest estates concurrently with mansions. William Coddington had a "larg Corne Barne" in which he housed twelve oxen, eight cows, and some other animals during the wintertime.*[9]

The silo in which Coddington's corn was stored would have been a chimney-shaped structure that occupied a corner inside the large barn. More commonly, what was referred to as a "silo" on early farms was "a large, shallow ditch, protected by thatch or briars from the rain, sun, and air, and used as a store-pit for potatoes, etc." The earliest in Rhode Island resembled stone-lined cellars, often built into the side of a hill, or at an angle to avoid direct sun light. A wood-framed "roof" was then constructed and covered with briars, hay or even corn stalks. In this respect, the "silo," before the advent of upright structures in the nineteenth century, would have been the first of a farm's outbuildings.

Many of these early farms were much smaller than those of the wealthier land owners, with acreages of "four to sixty acres" and barnyards that held a cluster of smaller outbuildings, or "appurtences; such as sheds, dairies, and other structures used for weaving or special purposes."

Typically, the outlay of an early farm would have the house alone on the highest point of land and set away from the road at the end of a long lane. The barn was usually at some distance from the house, while other outbuildings, such as a chicken coop or a smokehouse, would have been closer to the farmhouse in a clustered area that came to be known as the barnyard. Rhode Island communities were not typical due to the rocky hills, the swamplike meadows and thick woodlands that make up the landscape, much like the rest of neighboring New England. Historian and illustrator Eric Sloane wrote in *Our Vanishing Landscape* that "the prevailing wind, rain drainage, the contour of the land and proximity to the farmhouse decided where the ground cellar, the smoke house, the summer kitchen, the butchering shed, the woodshed, the spring house and the wash house went."[10]

The architectural form of a barn brought by early settlers to Rhode Island, as in neighboring colonies, would have been the English-style barn, which typically used three framed sections, or "bays," defined by vertical timbers. As with post-and-beam construction, heavy, hand-hewn timbers would have been joined and then trunneled together with wooden pegs before being braced. A sturdy, boxlike structure, the center bay served as "the drive" where carts were taken to be unloaded. This

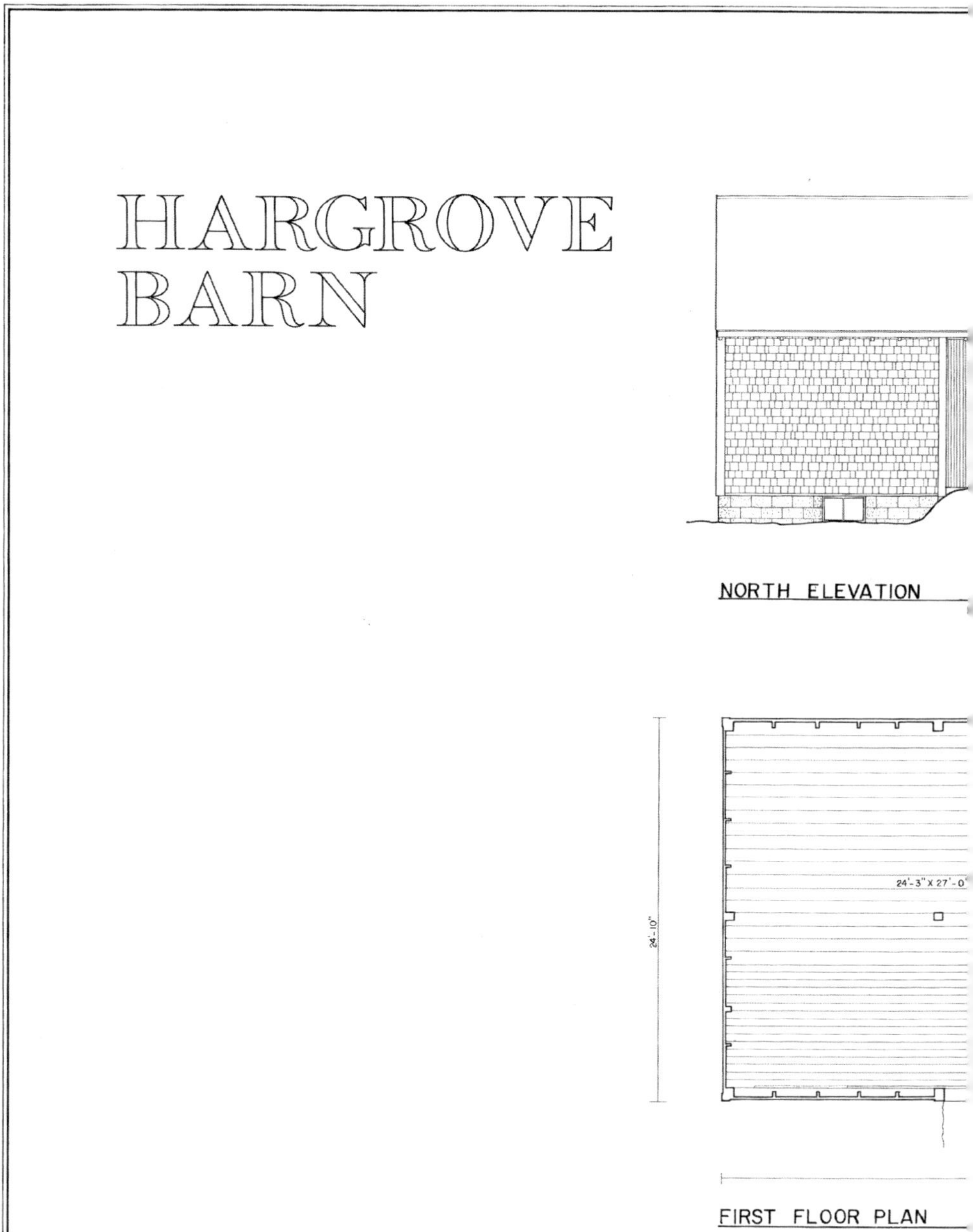

Diagram of a typical eighteenth-century American barn. *Courtesy of the Library of Congress.*

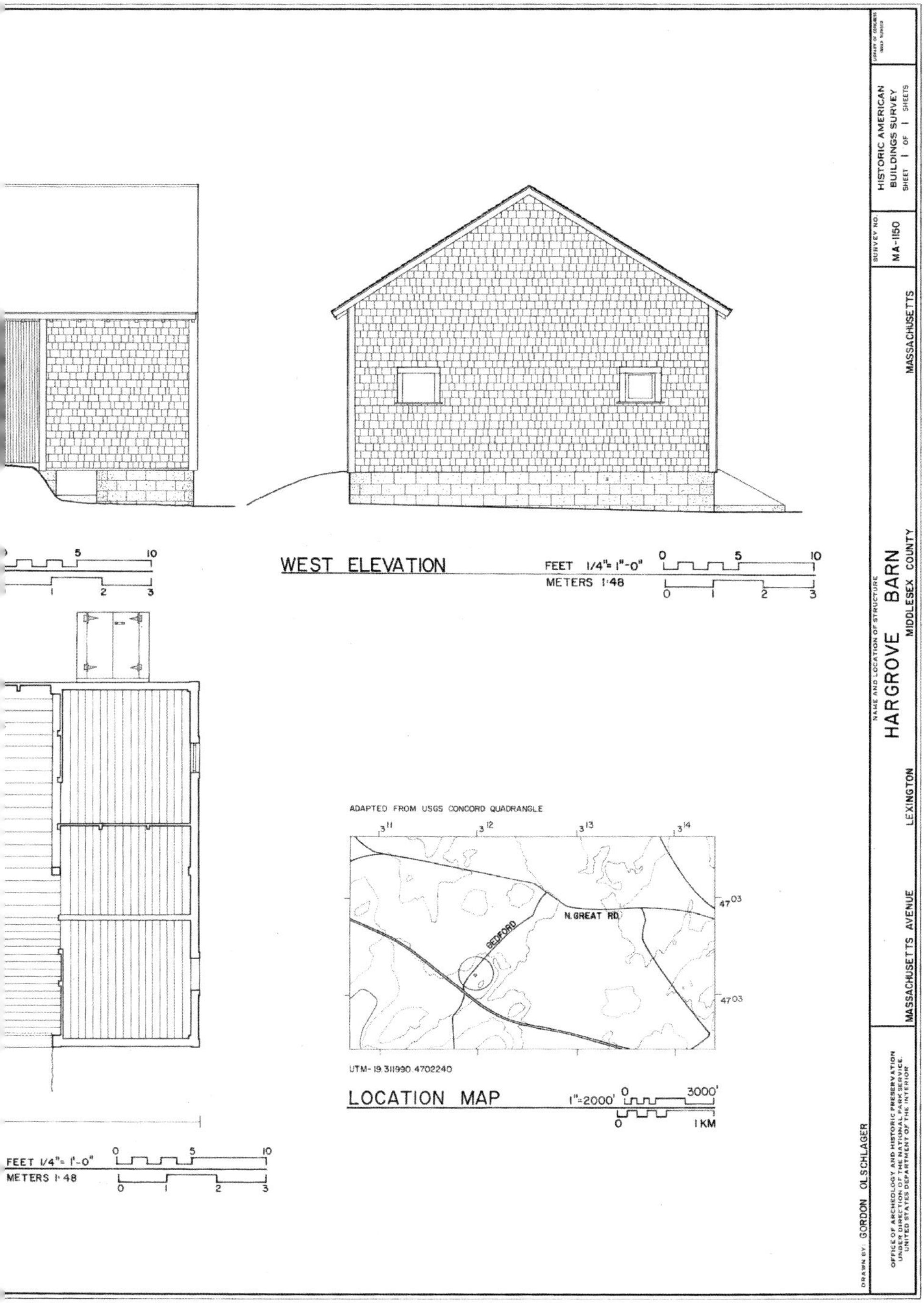
WEST ELEVATION
FEET 1/4"= 1'-0"
METERS 1:48
ADAPTED FROM USGS CONCORD QUADRANGLE
N. GREAT RD.
BEDFORD
UTM-19.311990.4702240
LOCATION MAP
1"=2000'
3000'
1 KM
FEET 1/4"= 1'-0"
METERS 1:48
DRAWN BY: GORDON OLSCHLAGER
OFFICE OF ARCHEOLOGY AND HISTORIC PRESERVATION
UNDER DIRECTION OF THE NATIONAL PARK SERVICE
UNITED STATES DEPARTMENT OF THE INTERIOR
NAME AND LOCATION OF STRUCTURE
HARGROVE BARN
MASSACHUSETTS AVENUE
LEXINGTON
MIDDLESEX COUNTY
MASSACHUSETTS
SURVEY NO.
MA-1150
HISTORIC AMERICAN
BUILDINGS SURVEY
SHEET 1 OF 1 SHEETS

Outbuildings at Coggeshell Farm, Bristol, Rhode Island. *Courtesy of the author.*

was where workers also shucked feed corn and threshed grain. The end bays or "mows" were used for storing grain and housing livestock.

A simple gable-roof design would have covered the structure, but there were no ridge rafters in these beginning barns; the rafters were crossed at the tips and trunneled with a mortise joint. If the pitch of the roof was not steep enough to allow snow to slide off in the winter, then the supporting struts inside were made all the stronger. The two methods most used were the "pitch-and-collar," a square-framed support within the loft, and the "great-strut," whose design stood great timbers braced at vertical angles against the frame of each side of the roof.[11]

There is evidence, according to Sloane, that much of the framework of New England's earliest barns was built with the timber from old boats. While there were plenty of pine and hardwoods, such as chestnut and hickory, in the woods around most farms, this seemed to be the only way to acquire the best well-seasoned oak after English scouts had marked the great trees inland from the coasts "with the King's broad arrow," to be cut and shipped back for use in building ships for the Royal Navy.

The American adaptations were to expand the size of the typical English barn, with the addition of a second floor for the storage of grain, leaving the ground level for the farm animals. They also changed the bracing on the frame, leaving behind the long-used "ships knees" of the English barns, and pegging a straight brace between the timbers.[12] Depending on the number of the herd and the builder's pocket, the barn might be a sizeable structure.

By the time of the English preacher George Fox's visit to the colony in 1672, both "a large barn…full of people" in Providence and a "justices barn" in Narragansett were available for him to preach in. Having been a shepherd as a boy, he noted the "great barns" that were larger than those in Essex, Devon or Lincolnshire. He also heard for the first time of the structure called a "cow barn."

For those curious about details in these early structures, the early barns generally had no glass windows.[13] Air and light were admitted through the opened wooden doors and sometimes through slatted louvers within the pine planking. Sometimes, an early barn had glass bottles "clayed" into a cut frame, but these were replaced with glass by the eighteenth century. Almost

An eighteenth-century barn at Wilbor Farm in Little Compton, Rhode Island. *Courtesy of the author.*

all early barns had a louvered or diamond-cut opening near the peak for ventilation that was called the "wind-eye," whose name evolved into the glass enclosed openings in houses and barns.

The doors of early barns held only wood-peg or cowhide hinges. The king's tax on metal hardware meant that "most barn doors swung on leather or wood hinges until the end of the eighteenth century."[14] Additionally, most early barns were without the ornamentation of a weather vane. If there was one above the roof, it was made of white pine and took the form of a fish, an arrow or a pointing hand. The metal roosters and horses were crafted after the end of the Revolutionary War, as were the hand-forged iron hinges strapped to doors of nineteenth-century barns.

Stone barns were also an import from European countries. One early such barn in Little Compton dates to the eighteenth century, and another still stands in Burrillville, built by an Englishman in 1855. The longevity of the barn, a "rubble work building, with a large barn door opening and several smaller doors, and a row of small, rectangular openings across the top,"[15] eventually gave the road along which it stands the name of Stone Barn Road.

One of the most prominent stone barns on the landscape was the Sheffield barn in Little Compton. A "two story high" dairy barn of "uncoursed random ashlar walls with openings for fenestration," this barn was recorded as being "the largest stone barn standing in the town" and of remarkable scale to other structures around the region.[16]

The large hip-gabled roof had collapsed in the 1970s, and today, only a portion of the walls remains. The great barn stands like an ancient ruin beside the ploughed fields of the working Moniz Farm on Stone Church Road. Barns of stone and timber from the eighteenth century are also located on West Main Road, as well as Peckham Road and on East Road on the route to Adamsville. A large stone-and-timber barn also rests on the Locust Valley Farm on Route 102 in Exeter. The nineteenth-century barn has now fallen into decay as a new complex sits across the highway.

Another iconic stone barn, though a more modern version, remains in Lincoln, Rhode Island, with the large barn built by Walter Theinert for his dairy business. He maintained the structure, as well as the stone walls that bordered the farm, for much of his life.

New England's rolling landscape led to another innovation of early American farmers: the building of "bank barns," structures that utilized a hillside to create part of the foundation for the wooden barn above. Stables and storage were located in the stone foundation built into the bank while

An eighteenth-century stone barn in Burriville. *Courtesy of the author.*

An eighteenth-century stone barn off West Main Road in Little Compton. *Courtesy of the author.*

the upper threshing floor was level with the crest of the hill, often by the roadside itself.

Of those necessary outbuildings on the farm, one of the most important would have been the corncrib, a shed where food was stored for pigs and cattle. The first were slatted sheds, lifted on posts and holding bins for both hard and soft corn. The corn was protected from rodents by nailing pie plates to the top of the posts underneath the shed.[17] Later innovations included tin and even glass plates affixed to the legs below the foundation.

On the larger farms, such as the Coddingtons' estate, massive corn barns were built to house both corn and cattle, but these were soon abandoned in favor of those smaller American designs, which allowed those with lots of acreage to establish several at key locations on a farm.

These corncribs, some of which still remain preserved on Rhode Island farms, were mostly of the earlier design, with the sidewalls always slanting out at the eaves; but now, they are shingled, as were most wooden barns by the nineteenth century.

Smokehouses were most often built of stone, as were dairies, in New England. Cider houses, chicken coops, sheds and privies were all constructed of wood, unless an owner of a large estate indulged his taste for more permanent structures. Icehouses, a New England tradition, as Eric Sloane wrote, had "no traditional style, so the farmer could express himself freely."

A farm most often had its own source of water power, either from a fast-flowing brook, a larger river or even a pond. These were essential to run the

"Bank Barn" in North Smithfield, Rhode Island. *Courtesy of the author.*

An eighteenth-century gristmill. *Photo by Harriet W. Kenney.*

mills needed on the property to provide clothing, timber, hand-cast hardware and especially, flour and cornmeal.

While there was usually a miller among the founders of a community who was given the task of grinding corn and making flour and meal at "a place of convenience" for the entire town, on a rural farm, with some distance between houses and even more from town, it became imperative to have several mills for differing purposes on the farm.

A fulling mill would likely have been the first to be built on a waterway, for use in making fleece into wool for clothing and blankets. With the predominance of sheep in Rhode Island, these not only provided a needed resource for the family but were also often a source of income as well. Sheep herding has had a long tradition on Rhode Island farms.

A descendant of an old Narragansett family recalled that among her earliest recollections was gazing from the old farmhouse's kitchen windows "facing the upland meadows, and woodland, where grazed a large flock of meek-eyed sheep. The wool from the sheep commanded a good price and was a considerable income for the home."[18]

The fulling process began in early September so that clothing could begin to be made by Thanksgiving.

The fleece would be brought to the mill and placed on tenterhooks, stretched out in a tub of water and detergents and then pounded by water-powered wooden mallets, often for days, until the oils were removed and the fleece beaten into a workable felt cloth.

Where the water was off shore, early farms built windmills to harness the sea-breezes and power the millstones for grinding corn and flour. These, and other mills over time, would be adapted for different uses. By the late nineteenth century, there were about twenty windmills operating in Rhode Island. While a windmill gave the miller the advantage of year-round use, those who ground corn and flour by windmill faced dangerous work, and there were many fatalities. A miller mending the sailcloth on a vane could be thrown off by a sudden gust of wind, struck by a spar or, most deadly, caught in the wheels and mangled by the gears. A millstone that had caused a death was considered unlucky and discarded after such an accident. Sometimes, the stone was used as the grave marker for the unfortunate miller.[19]

While the early agricultural and technological growth in the colony suffered a setback during the conflict of King Philip's war, when nearly every farmhouse, barn and outbuilding between Providence and Point Judith was burned to the ground,[20] Rhode Island would make a "remarkable recovery." Within two years, the cattle trade was making a comeback, as was North Kingston, a farming town that had established itself just a year before the conflict began. In 1677, five thousand acres of land was laid out in one-hundred-acre lots for the town of East Greenwich, a new settlement made up of men from the now burgeoning farmlands of Portsmouth and Newport.

Bridenbaugh wrote:

> *The unqualified success of agriculture in Rhode Island and some settlements on the mainland throughout the 17th century resulted not so much from the collective effort of many pioneer families, each toiling away in a few cleared*

> *acres...as from the accomplishments of a few gentlemen who, investing large sums of money for that time and place, created in New England a species of landed estates superficially resembling those of Old England, but unique in the colonies.*[21]

These would be the first of the "Narragansett Planters," and a description of these families bound by marriage, their grand estates and the lifestyle lived among them by the turn of the century can be found in the biography of "College Tom"—that is, Thomas Hazard, the fourth generation of his family on the Hazard estate in Narragansett:

> *At the beginning of the 18th century the country was taken up by great farmers working their farms with slave labor, either indian, or negro, or both, and living the life of English squires...their books and their tea came from England direct to Newport...Up from Point Judith, through the bridal path that led from one great farm to another, divided by stone walls and heavy gates, came the ladies in their camblitt cloaks, and the gentlemen in broadcloth and britches, with silver shoe and knee buckles, mounted on the Narragansett pacers of famous memory...From Boston Neck the gentry gathered, from Little Rest, and the farms of Matunuc.*[22]

Robert Hazard was the patriarch of the largest and most influential of the planter families. He was paying tax as a landowner in "Kings Town" as early as 1687, and before his death, he would increase his land holdings to a large parcel on Point Judith Neck, which he deeded to his son Stephen. It was Hazard's oldest son, Thomas, however, who would greatly increase the family holdings by purchasing over nine hundred acres from Samuell Sewal, the son-in-law and heir of John Hull. Three hundred of these acres lay on "the west side of the Saugutucket River, near the village now known as Peacedale" and the remainder "on the Back Side of the ponds," being west of Matunuck. He also acquired land on Point Judith Neck and later purchased the southern end of Boston Neck. Thomas would divide much of these lands between his sons before his death, including a "great cattle range" his son Robert inherited by Worden's Pond.

The Robinsons were another prominent family of Narragansett. Rowland Robinson, having lived in Newport since 1675, purchased three hundred acres "east by the salt water, west by Petticomcitt pond" on Boston Neck. Robinson built a farm on this land and acquired other parcels, including the purchase of a three-thousand acre lot from the "vacant lands" ordered to be sold by the colony. This tract was adjacent to the Connecticut border, and Robinson referred to it as "my Wood River farm."

His son William increased the family's landholdings even further by acquiring large tracts of land from the heirs of William Brenton, including "Silver Lane" and "Sugar Loaf Hill." He also purchased a parcel on Point Judith Neck from William Mumford and continued to maintain the family farm.

The Gardners were also early settlers in the area, beginning with the purchase of one thousand acres of land by George Gardner in 1663. Gardner raised fourteen children on his farm, many of whom would become prominent citizens of Narragansett in their own right as the family continued to prosper. By the time of his grandson William's death, the family held title to 1,600 acres.

The Coles, Babcocks, Potters, Reynolds, Helmes, Willits, Carpenters, Congdons and Watsons, as well as the Updikes in North Kingston and the Champlins and Stantons in Charlestown, became prominent families among the planters. As historian Christian McBurney noted in his *History of North Kingston*:

Hauling timber for the farms in Narragansett. *Courtesy of the Pettasquamitt Historical Society.*

Casey Farm, circa 1750. This landmark house along Narragansett Bay was built by Daniel Coggeshall on the tract of land first owned by William Brenton. The barn and outbuildings have been attributed to the time of Thomas Lincoln Casey's management of the farm in the mid-nineteenth century. *Courtesy of Historic New England.*

> *By 1730, successful Narragansett Planters...owned thousands of acres of land. The average...planter held about 400 sheep, 80 cattle, and 20 horses.*[23]

Several planters attempted to grow English grains, but these fared poorly. Wheat, barley and oats were susceptible to disease promoted by the "frequent fogs during the spring and summer."[24] Rye apparently grew well, as did flax. Some planters, Thomas Potter among them, attempted to raise a large cash crop of tobacco from the soil, and this would have necessitated the building of "tobacco-barns." Little records exist in the colony, but from neighboring farms, we know that these long, airy barns were built with three levels of beams, from which leaves could be hung to dry, and opened slats on all four walls.

An example of a planter's estate remains one of the most well-preserved farms in Rhode Island. Casey Farm, as it is now called, was originally established by Daniel Coggeshall and his wife, Mary Wanton Coggeshall, in 1750.

The farm, like that of other planters, had a substantial herd of cows for producing cheese, and the family grew and exported barley and apples to neighboring Newport as well.

Many of the planters built docks that stretched from their land to open water, and the proximity of Newport ensured that these goods—as well as timber, cheese, stock cattle, sheep and horses—could be shipped to the southern colonies and ports in the Caribbean.

Inland, and farther up the coast from these southern shores of Narragansett Bay, smaller farms were populating the countryside. When the town of East Greenwich was incorporated, ninety-acre lots were granted to the first settlers, including John Spencer, Clement Weaver, Charles Macarty and Giles Pearce, in whose home the first town meeting was held. When he died in 1698, his will bore testament to the the humbler, but still significant farms that were more common than the large estates of the planters.

Pearce left his son, Jeremiah, ninety acres of land, a pair of oxen, three cows and a horse. To his daughter, Susanna, he bequeathed "a feather bed, thirty pounds, two cows, and a heifer."[25] An inventory of his estate shows that among Pearce's property were

> *four oxen, seven cows, several heifers, four calves, eighteen sheep, eight lambs, a horse, and a mare.*

The farmer also owned a slave girl named Frances, who in his will, was stipulated to remain with his wife until her death, whereby she would become the possession of daughter Susanna.

Thomas Fry Jr. inherited the original land grant from his father, a general sergeant in Newport, and soon expanded his holdings to include land along Middle Road, South County Trail and the Frenchtown section of East Greenwich. Among his descendants were Captain Benjamin Fry, who ran the farm like a small plantation, importing slaves for both the farm and for profit elsewhere. As early as 1719, the Fry Farm was producing dairy products and "bartering" with neighbors to extend credit for accounts settled later. It has remained a working farm ever since. In time, nine generations remained on the original farm along South County Trail.

Providence settler and surgeon John Greene was the first to have a substantial farm in what became known as the town of Warwick. His purchase in 1642 of 660 acres of land extended from Shawomet to the cove named Occupassuatuxet and into the bay itself, with a small 14-acre island on which he built an early iron forge.

The farm, which became known as "Spring Green," was the family's home for over 140 years, after which it was purchased by wealthy merchant John Brown of Providence in 1782 for the unheard of sum of $3,000 in silver. The large, gable-roofed house that Greene's descendant built still stands on the property, and the Greene family had, by that time, expanded to farms elsewhere in Warwick. John Brown built a large complex of barns on the property beginning around 1788. His descendant John Brown Francis built a stable that, by 1862, held seventeen horses. The barns remained until September 1923, when they were consumed by fire.

The Browns used the estate as a summer home but utilized the farm year-round and leased much of the land, as John Greene had, to local farmers. Spring Green was also home to one of the first icehouses built in New England. John Brown had the conical, wood-shingled structure built in 1786 beside a pond on the property with an ice run for easy hauling of the cut blocks of ice to the house.

John Greene's son James established a farm later called Forge Farm as early as 1684, and the farmhouse that still stands reputedly contains part of the original structure. James's son Jabez Greene dammed the Hunt River and built a sawmill and gristmill in partnership with Timothy Hill, his neighbor across the river. The farmhouse was enlarged in the mid-

An icehouse built by John Brown on the "Greene's Holds" property. *Courtesy of Henry A.L. Brown.*

eighteenth century, likely around the time of General Nathanial Greene's birth in the home in 1742.

The Greene-Bowen Farm in Nassauket was established by another James Greene, the nephew of surgeon John Greene, also around 1684, who built a house along what is now Buttonwoods Avenue sometime between then and 1715.[26] The original farmhouse was a one-and-a-half-story structure with a brick-end chimney, rather than the traditional stone-ender. The house was also "added to soon after its completion" with "a unique amalgam of features found in early Newport houses and features more typically found in Providence buildings."[27] Inside, the main room has a large fireplace and timbered mantelpiece. A locked watch window is built into the large post to the right of the fireplace. In the dining room next door, a slightly smaller fireplace opens to the room.[28]

James Greene's farm consisted of 140 acres, the house, a barn, a cheese house and an outhouse. The farm was later owned by a son named Fones Greene. At the time of the 1779 census, the acreage was divided as follows: 90 acres of pasture, 18 acres of meadow, 13 acres of woodland, 8 acres of tilled fields, 4 acres of orchards and 1 acre of tobacco. The farm remained in the Greene family until 1900, when it was sold to the Bowens.[29]

The Budlong Farm, established around 1700 by John Budlong, also contains an early farmhouse. The framing in the southeast room of the present structure suggests it was the original house—a one-room, chimney-end dwelling. The farm remained in the Budlong Family until the death of Henry W. Budlong. While owner of the farm, Henry ran a campground for working-class families on his property adjacent to Greenwich Bay. After his death, the farm was left to the housekeeper's daughter, Emily Ruville Hohler. Mrs. Hohler and her husband ran the farm as a dairy and made some alterations to the farmhouse. By the 1980s, her descendants were running a horse stable on the farm.

Another early farm was at the site of the Eli Randall House built in 1701. For many years, a millstone belonging to Locust Farm remained propped up outside the house, and a small miller's house lies in the rear of the property, now surrounded by modern ranch-style dwellings on Sandy Lane.

Benoi Waterman acquired the Waterman homestead in old Warwick near what would later become Wharf Street beside Apponaug Cove. He kept a general store and his self-sustaining farm, which provided sheep and timber for the community, as well as supplies for his store. Waterman often leased his pastures for neighbors' cattle and horses, billing John Lippit for "pasturing your horse 8 weeks" in July 1737. He brewed beer and cider as well, charging customers by the "barrill." Benoi's brother Resolve established a farm in what is now Greenville, which was self-sustaining as well and which had a large orchard, whose cider supplied the tavern he also operated on Putnam Pike, less than a mile from his farm on Austin Avenue. The farm would survive for generations, becoming known as "Maplewood" when the owner during the Civil War planted a stand of maple sugar trees on the property that produced syrup for many years.

In neighboring Scituate, there were a number of early farms, especially after 1700. Early settlers included the Angells, Aldriches, Arnolds, Browns, Harrises, Hopkinses, Mathewsons, Smiths and Wilkinsons. For several generations, these families worked the land and produced sons and daughters who in turn established their own farms and families. The Angell Tavern was established and served travelers on Plainfield Pike and was likely the first business in town. Sawmills, gristmills and forges were also established shortly after the town was incorporated in 1731. The Brown Homestead on what is now Rocky Hill Road is an example of these early farms. A two-and-a-half-story building with a large central chimney, the house is distinguished by a wing and piazza, as well as an added portico entry from the nineteenth century. A fine wood-shingled barn and woodshed stood on the property well into the twentieth century. The homestead on Gleaner Chapel Road, later called the Martin Smith Farm, also holds a house from the period. A two-and-a-half-story structure, end to road, holds two brick interior chimneys and a central entry with transom lights in a four-bay façade with a one-and-a-half-story wing at the rear. Israel Smith's family purchased the farm in 1785, and the estate remained in the family for nearly two hundred years. The farm was eventually named for Martin Smith, Israel's grandson, who rose to prominence during the Civil War and served in the assembly and town council after settling on the family estate in 1876. After the farm was sold by the family in 1958, the new owners restored the house and sold several outbuildings to be preserved at Old Sturbridge Village and the Guillford (Connecticut) Keeping Society.[30]

After 1800, when the population of the town had reached over 2,500 people, larger farms began to be established that sold produce locally, as well as in Providence. Chopmist Hill Road holds a few examples of farms from this era, including the Ridge View Farm, whose original one-and-a-half-story house dates from 1790 and is now a rear ell for the two-and-a-half-story addition built in 1850. The farm is outlined by dry stone

The barn complex at Spring Green Farm, circa 1923. The center of the barn complex was the first to be raised in June 1788. *Courtesy of Henry A.L. Brown.*

walls, and a family cemetery sits south of the house. The Hopkins Farm holds a classic Federal farmhouse constructed in 1812 by Augustus and Carver Hopkins on a 143-acre tract. The farm was a working farm until 1859. The Dexter Arnold Farm (1813) also holds a Federal, two-and-a-half-story house with a central chimney. The central entry holds transom lights and paneled pilasters in its five-bay façade. The farm has been entered in the National Register since the 1970s and still holds some fine examples of outbuildings with a combined woodshed, carpenter shop and chicken coop, as well as a corncrib and privy on the property. At its height of productivity in the nineteenth century, the farm's crops included corn, barley, beans, potatoes and dairy products for transport to Providence as well as local sales.

In rural villages such as Coventry, Greene and West Greenwich, timber was a prized commodity and a source of income for early settlers. A historian of the town of Greene wrote:

> *The main road south from Greene to Hopkins Hollow, beyond into West Greenwich and eastward toward Summit...was much used by the heavy team wagons driving wood and lumber or logs for the mill here. Anywhere from twenty to thirty teams, using from two to four horses, were sent over this road, usually making two trips a day, one in*

the forenoon, another in the afternoon, bringing their loads to Greene, and returning sometimes empty and sometimes loaded with grain to be used at home.

As with other villages, there were farms dotted throughout the countryside well before the towns became officially incorporated. One such early farm was the Woodlot Farm (1737), a small, self-sustaining farm located nearby Carr's Pond. The simple farmhouse has a steep-pitched roof and was restored late in the twentieth century by artist Maxwell Mays, whose family bequeathed the farm as a nature preserve for the town.

Larger farms also existed, including the Hill Farm, established as early as 1741 on two hundred acres of land. The farm became a manufacturing site by the mid-1800s, with a japannery and shoelace factory located on the premises, which required a second two-story house to be built on the property. The Hill Farm extended as far as Johnson's Pond, where the family became the first to built waterside resort cottages in the 1920s, leaving the surrounding fields to be used for growing feed corn.

The Broadwall Farm (1800) held expansive acreage along Sisson Road, and many of the picturesque stone walls that still line the road are remnants of the farm. The handsome two-and-a-half-story gable-roofed house with a central chimney and a one-story ell still stands and remains beautifully maintained. The farm in its current operation sells Angus-Hereford cross calves and heifers as well as breeding stock.

One of the most impressive farmhouses from this early period is on the property of the former Windy Parks Farm, established by Israel Wilson in 1814. One of the wealthiest farmers in the area, Wilson built a two-and-a-half-story Federal house with a formal pedimented doorway flanked by fluted pilasters and capped by a semicircular fanlight with tracery. Wilson's land extended across from the farmhouse on Harkney Hill Road and included Quidneck Pond, where he operated a fulling mill, sawmill and gristmill before selling the pond to the Sprague family for use as a reservoir in 1846. Much of the land would later be incorporated into the Knight Farm by the end of the century. The Arnold Farm was established on Narrow Lane sometime before 1838 by Nathaniel and Lydia Vaughn Arnold, when the present, sprawling farmhouse was originally constructed. Their son, Edward Everett Arnold, born in 1853, increased the farms holdings by a considerable degree. He was educated and became a successful businessman in Providence. Over the years, he bought up adjacent property as he "summered" on the farm and improved his property. The farm remained in the family for over a century and was later the site of the Greene Herb Farm (1942–72), an early predecessor of a popular use of farmland today.

By far, the largest farm in the vicinity would have been the Nicholas Farm, established in 1813. At one time, the farm contained more than one thousand acres and extended into Connecticut. It was so large and isolated that the children of the farm workers were granted their own schoolhouse, which served well into the twentieth century.[31]

North of Providence, in the area listed as the "north woods" on early maps, William Blackstone had established a farm at a place he named "Study Hill" in 1635, where he planted the first apple orchard in the colony and became known for traveling far and wide on a tamed white bull. There is some speculation as to whether he may have held the first herd of milking cows in the colony. Later farms in the area began to be established in 1658, when Arthur Fenner, Roger Mowry, John Sayles and Valentine Whitman were empowered by Providence to "treat with the Indians that lay claim to the meadow of Lokussuk and clear it for the town."[32]

In 1660, John Whipple Sr. had lands laid out to him, and he later acquired more property, which was divided among his sons: Eliazer, William and Samuel. A second farmhouse was built around 1680, and this was still standing one hundred years ago, as was the Valentine Whitman House just north of the Whipple farmlands. To the west of Whipple's farm lay that of his son-in-law, John Wilkinson. Both these men were wounded "almost within sight of their own doors" with the outbreak of King Philip's war.

To the south of these farms lay those of William Whipple and John Dexter, and farther south eventually lay the farm adjacent to the "end of the world," a long pond which acquired the name of the Colonel Sylvanus Scott a few generations later.

In 1661, in the adjacent area known as Lime Rock, Thomas Arnold purchased a substantial tract of land where he established a farm. It remained in the family for generations and became a dairy farm in 1860, when the current house, barns and outbuildings on the property were built.

Just southwest of Lime Rock, in 1696, Elisha Smith, grandson of the Providence miller, had established his farm on the northern edge of a large body of water connected to the Stillwater reservoir by a rocky stream. He built the farmhouse, barn, mill and other outbuildings early on, and then in 1730, the size of the house was doubled with the addition of a two-room house that was purchased by Smith and brought by oxen to the farm.

John Mawry and Edward Inman purchased of two thousand acres in the year 1661 from "William Minnian" of Massachusetts Bay. Minnian was the English name for the sachem Quashawannamut of the Massachusetts tribe, and as the naming of this Native American suggests, the English were by then well established in the region. The "Great Road" that Inman built his "plantation" on was a well-traveled route that connected Providence to Worcester and Boston.

In June 1675, Inman and seven others signed an agreement that established the second metals-mining operation in the colony.[33]

But Inman was forced to fight for his land time and again. Initially, Providence tried to claim rights to it, until a Newport court in 1672 ratified "all and every part" of the deed. Despite this, as more settlers established farms after the conflict of King Philip's war, Providence formed a committee to meet with Inman and "debate the matter with him about the lands where on hee and other[s] with him are settled upon"[34]

Eventually, an agreement was reached on the boundaries of "Wesquadomsett" and its 3,500 acres, the largest of these bordering the "Pawutcucket"(Blackstone) River and "Wasquadomsett" River (Crookfall Brook) and extending west and south through the "great Cedar Swamp" and into Nipsachuck, an area of ancient Native American ceremonial sites now under federal protection.

Along these waterways from Crookfall Brook and the manmade reservoirs that altered the river's natural course, one finds foundations of mills, stables and houses. Most striking of these remains is the mile-and-a-quarter long "road" built entirely of stone that allowed carts to run from the waterside mill, through the swampland to the barn located in a meadow beyond the woodlands.

This mill site was attached to one of the larger farms in the area; others were established on a smaller scale and were family-run operations. The Inmans had a sawmill established at the mouth of the Tarklin River and a forge on the Branch River, as well as a gristmill situated a mile west of what was later named Slatersville.

Other families built sawmills, fulling mills, tanning yards, sawmills and gristmills along the numerous small watercourses that run through Woonsocket and North Smithfield. Limestone quarries and processing mills mushroomed in the community as well, leading to one village being named "Lime Rock" for the export of the product to Providence and beyond.[35]

The cart road built from the mill foundation beside Woonsocket Reservoir number three. *Courtesy of the author.*

Another eighteenth century sawmill was reputed to be along Cherry Brook, and yet another was set up along the Woonasquatucket River, near today's Primrose Pond. By 1795, Elisha Bartlett, who had once worked for Inman's forge, was manufacturing his own scythes and edge tools at his forge along the Branch River.

Outside these established communities, in the far-flung rural reaches of the smallest colony, other settlers like Blackstone established early farms as well. Samuel Wilbor left Portsmouth in the 1680s and settled in Little Compton with his bride, Mary Potter, in a two-room house he built off the road that connected the town to New Bedford, Massachusetts. The Wilbor House, as it is still known, was initially home to eleven children. It grew into a ten-room farmhouse and remained the family home for 250 years.

Now owned by the Little Compton Historical Society, many of the outbuildings have been preserved, including a corncrib that dates from the early eighteenth century, a wooden vertical board–sided barn from the same era and a later wooden-shingled barn built in 1850.

As with other communities, Little Compton was the home of many small, self-sustaining farms in the eighteenth century, whose surplus of produce, or stock, was likely sold at stores in Tiverton, Adamsville, and New Bedford. According to the survey conducted by the Rhode Island Historic Preservation Commission, what distinguished the small community from others in the region was that, in large part due to its relative isolation, Little Compton retained its agricultural integrity far longer than many in the region. "For nearly two hundred years, from the beginning of the eighteenth century until the end of the nineteenth, the town's agricultural production and its population changed remarkably little in comparison with the rest of the state."

Along the Great Road and with the building of the Louisquisset Pike, Douglas Pike, Putnam Pike and other routes, large farms began to populate the rolling hillsides of rural Rhode Island.

Some of the great early farms in North Smithfield include the Aldrich Farm on Comstock Road, which retained its vertical board–sided barn, corncrib and outbuildings. The Smith Farm on Grange Road held an early barn and outbuildings until recent development carved up the farm.

The town of Glocester, settled in 1706, became one of the largest and most well-known of farming communities, with early farms along Putnam Pike clearing much of the land for agriculture and grazing. The farms grew corn, barley and rye, as well as a bounty of vegetables. Many had orchards as well.

An eighteenth- to nineteenth-century barn on the property of the Smith-Appleby house. *Courtesy of the author.*

Sheep and cattle grazed in the open pastures and seemingly grew in such great numbers that the town deemed it necessary to build a pound for stray beasts in 1746 on Chopmist Hill Road. Early farms include the William Steere Farm on Long Entry Road, the Colwell Farm on Tarklin Road and the Place Farm off Snake Hill Road. These farms contain early unpainted vertical board–sided barns.

Other early farms with wood-shingled barns and outbuildings include the Smith Farm, the Thomas Cutler Farm and the Hunt-Farnum Farm on Putnam Pike. Those farms established off the main road include the Reynolds Farm on Snake Hill Road and the C.C. Mathewson Farm on Tourtellot Hill Road. Many of these farms and their early barns and outbuildings remained intact into the twentieth century. An article on early motor travel in the area exclaimed of Glocester: "One of the most striking features to impress a stranger driving through the archaic township is the large number of great barns standing close to the highway."

An eighteenth-century barn on former Steere Farm, circa 1760, in Glocester, Rhode Island. *Courtesy of the author.*

In the years leading up to the American Revolution, the farms that grew in the colony were predominantly dairy farms, and so great barns began to appear on the landscape, an icon of an industry that was to have its own long tradition in the state. The New England Barns—or "Yankee Barns," as they have long been called—were long, airy barns with the drive running from gable to gable with stanchions on one side and hay storage on the other. The largest barns were bilevel, with stanchions several feet below grade level on both sides and expanded storage feed above. If the barn needed to be expanded, another bay was simply added at a gable end.

An early barn, still located outside Chepachet off the old Putnam Pike was constructed as a twenty-by-twenty-foot barn in 1680. Over three hundred years later, the barn measures sixty by twenty feet, with two additions having been added in 1820 and again in 1880.[36]

A plan for "the New England Barn" in William Pain's *The Carpenter's Pocket Dictionary Containing the Best Methods of Framing Timber Buildings* explains the basic design:

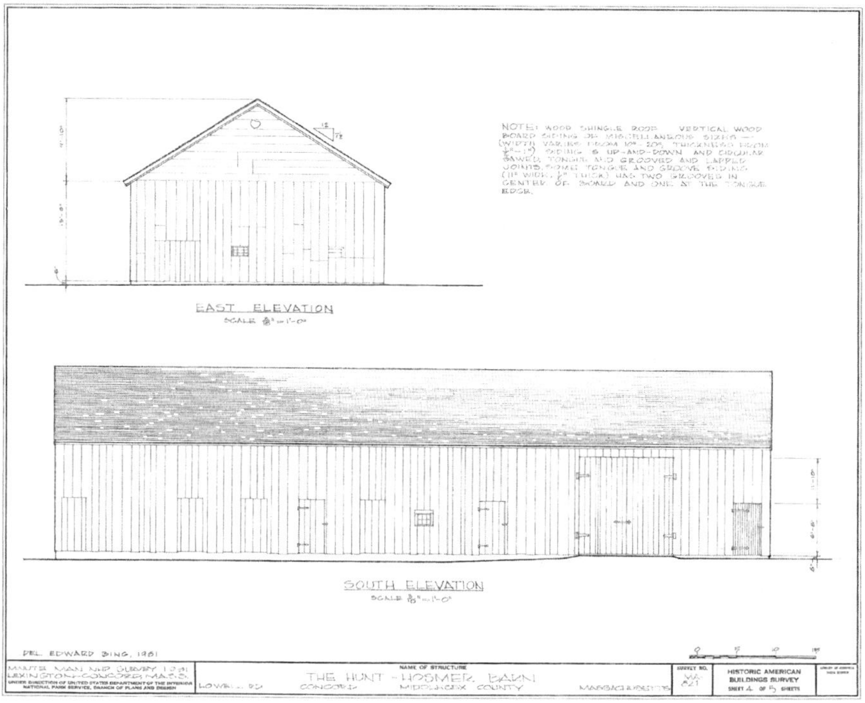

Design of a New England dairy barn drawn for the Historical Archives, 1926. *Courtesy of the Library of Congress.*

> *The whole width from out to out is 28 feet, and 26 feet 4 inches within, with gable ends for more convenient room…the ground sills need to be 10 inches by 6, the main post 8 inches by 10, the door post 8 inches square.*

The book lists precise measurements in its "table of timbers" for the interstices; quarters; braces from post to beam; teffel, or "top part of the post"; buttresses; beams; king-post; rafters; and principle rafters, which "at bottom must be 9 inches by 5, at to 7 inches by 5."

These dairy and agricultural farms produced goods for local markets and also often used their houses as "ordinaries," or places licensed to provide housing and meals for travelers. These farms often had large horse barns as well, not only to provide rest and oats for the individual animals but also to supply fresh horses for the coaches that now traveled the routes regularly.

Union village, originally called "Woonsocket," prior to 1800 had but a handful of eighteenth-century farms and Peleg Arnold's tavern strung out along the Great Road, but it soon expanded in response to the increased traffic at this crossroads of both the north–south route connecting Providence and Worcester and the east–west route that connected Boston to eastern Connecticut.

Builder Walter Allen attached his own two-and-a-half-story dwelling in 1802 to the small house built a century before by Hazadiah Comstock. He also built the Seth Allen tavern just up Great Road, as well as the Aldrich House and Tavern just across the road and the Stephen Brownell House with its great carriage house on the property, all in the space of a decade.

Closer to Providence, the need for stables and lodging grew as well. Around 1800, Nate Mowry built a house, barn and large stable along the family farmlands and opened the Nate Mowry Inn. The house held a large taproom and an expansive porch. The stable was ample enough to provide fresh horses, and the tavern became well known as a horse-changing stop for the coaches that traveled between Woonsocket and Providence.

Peleg Arnold Tavern on Great Road and Woonsocket Hill Road in Union Village. *Courtesy of the author.*

The carriage house behind the Walter Allen House in Union Village. *Courtesy of the author.*

Most fittingly today, the house remains, and the old horse stables have been replaced by modern stables and an indoor equestrian pacing ring. The Nate Mowry Inn, its barn and outbuildings are now part of Ballard Farm. A few of the old outbuildings are still in use, and the large hay barn, dated around 1810, has been restored by the Filippe family, who has owned the property since the 1970s.

Paul Filippe was kind enough to show me the interior of the hay barn, whose hand-hewn post-and-beam construction, including fitted and pegged timbers in the frame, is still sturdy today. The original barn board was used to panel the interior and create doors within the structure. Paul explained that there were four or five stalls in the ground chamber of the barn, and upstairs was the hayloft.

The restored hay loft holds the "winter office" for Ballard's restaurant, which the family owns on Block Island. The detail of work on the timber frames is a tribute to the craftsmanship that built the barn. Simple as it was for its purposes, it was hand hewn and built to last.

The horse barns on the property date from the 1970s and the 1990s as the family expanded the stables and built the inside ring for young riders to practice their equestrian dressage runs in the many horse shows that enthusiasts compete in throughout New England.

A restored hay barn, built circa 1810, at the Nate Mowry Inn, now Ballard's Farm. *Courtesy of the author.*

Interior of the hay barn on Ballards Farm. Note the hay vice that pulled bales into the barn through the shuttered opening where the window is now placed. *Courtesy of the author.*

Close-up of the joined and pegged timbers in the post-and-beam frame. *Photo by James Allen.*

In Providence by 1775, husbandry had adapted in an orderly integration with river commerce and industry. Those who maintained self-sustaining farms within the city were bound by ordinance to house and fence their domestic livestock well.

Most of the farms still used the agricultural lands on what still remained of Weybosset meadows and brought their mown hay to the scale "on the Town Parade," which had been for some years at what would later be called Market Square.[37]

Some earlier settlers of Providence purchased land elsewhere, leaving their small farms in Providence, or other lands under lease to tenant farmers. One such early case was that of John Steere, who wrote in the court documents:

> *An Indian by the name of Sam Noforce who hath for some years Lived by mee and hath well behaved himself toward mee and mine: A Certain Parcel of Land Laid out to mee for six acres beit more or less: for him the said Sam Noforce and his heirs for the full term of thirty years from the date of these presents...during the said thirty years the said Noforce shall have the sole management and profits of the said lands: to plant to corn or fruit trees.*

Most of the farms were small, self-sustaining establishments. Even the founder's families, by the second and third generation, were still living by fairly modest means. The inventory of Deacon James, who had married into the Angell family, shows that the homestead he inherited was a "substantial dwelling" with four or five rooms and a lean-to addition. The property included a barn and another outbuilding, which was likely a workshop. The inventory of the barn shows "a Horse, two Coults, a yoke of Oxen and two Cows." Also listed in "Ye Barne" were "a cart and shodd wheeles, some carpenters tools, narrow axes and one broad axe…a beetle and wedges, a sithe and two bells, a Tumbrill, a wheele barrow & two sleds, one swine, 1 hundred pounds of Tarro, and one Loome & gears & one pair of woosted combs."[38]

The homestead of John Field, part of which would later be named "Field's Point," was located on "the western side of the salt water "in the neighborhood of Cowpen Point, which jutted out into the Providence River near the junction of Point and Eddy streets. Field also owned a tract of land on "Weybosset Plains," as well as "the path to Mashapaug running

John Jenkes barn, circa 1714, on Benefit Street, Providence. *Photo courtesy of the author.*

through it." Mashpaug was a tract of meadow that lay a mile or so southeast of Mashapaug pond, which was an original boundary of the town. Field's farm and lands later inherited by his sons provided fresh produce and dairy products for the city until the mid-nineteenth century, when what wasn't sold of the large estate became a city park.

Other farms in the outlying regions of the city developed after 1710, along the completed Plainfield Pike. These supplied Providence with goods for many years. One early farm was the King Farm, at the foot of Neutaconkanut Hill, with its barn attached to the house. This form was uncommon in Rhode Island, though it had its use in neighboring colonies until town councils enforced laws against such structures because of fire.

Connected house and barn. *Photo by Harriet W. Kenney.*

The King farmhouse with attached barn on Plainfield Street. *Photograph by J. Cady, Providence.*

The King Farm began around 1720, and a descendant of the family became governor and claimed the entirety of the ancient hill used by the Narragansett people as a natural base for furnaces used in pottery and tool making. Governor Samuel Ward King built a massive estate on the hill and paved roads that wound their way from the grand cast-iron gate on Plainfield Street upward to the great house that overshadowed his ancestors' humble farm below.

The descendants of Thomas Angell of Providence were also among the early settlers of the area that would become incorporated as the town of Johnston in 1759. Daniel Angell established a farm on what is now Dean Avenue, and another Angell Farm was established on Putnam Pike in 1768. Other early farms included the Winsor Farm on Sikkibunkiat Hill, whose outbuildings long revealed the evolution of the farm from the seventeenth to the eighteenth century.

Looking at a photograph from the 1980s, we can still see in the rear of the farm an old structure with the long, sloping roof of seventeenth-century design. It was built on the slight bank of a hill, allowing its stone

The remaining nineteenth-century outbuilding on the Winsor Farm in Scituate, Rhode Island. *Courtesy of the author.*

foundation to be elevated slightly in the rear, enough for a cart and wagon. It featured two entryways, side by side, with a window between and a hay door directly above.

An eighteenth-century barn lay just to the right of this structure. The barn had wide wooden doors and a smaller door to the right. Three small windows, spaced with the doors between them, faced the slightly elevated earthen ramp leading to the barn doors. There were two larger windows on the side of the barn with four slatted vents on either side of the windows and another pair at the peak of the roof, which held a cupola. There was a narrow vent above the portal of the entry doors.

In Providence by 1750, there were several farms remaining within the boundaries of the city. Many were connected to the Brown family. One farm was the dairy established by Robert Brown in 1732 in the "wilderness" outside the cultivated city, close by the area known as "Cat Swamp" along the Seekonk River.

Jeremiah Dexter established his farm on lands that had been in the family since his great-grandfather's friendship with Roger Williams, receiving the last of the original home grants deeded in Providence. Jeremiah built a barn, corncrib and tapped a well on the land that was to supply the troops of Compte de Rochambeau when they encamped on the farm during the Revolutionary War.

The city was transformed in the latter eighteenth century from a quiet, backwater settlement to an active port and aggressive competitor in the China trade with merchants from the neighboring cities of Boston, Salem and Newport. Elegant houses, equal to those of the planters' estates in Narragansett, began to be built in the city, beginning with John Brown's elegant brick Georgian estate in 1786–1788.

Watchmaker Seril Dodge built a pair of elegant brick houses on land he purchased from his friend Moses Brown along steep Angell Lane in 1789. He removed a barn off his newly acquired land to another parcel of Brown property in order to build a cobbled gangway that led between the houses to a courtyard and a grand "Chaise House" in the northwest corner of the property. An "old stable" still stood in the northeastern corner, close to

Detail of the Sullivan-Dorr House stables in Providence, Rhode Island. *Courtesy of the author.*

Benefit Street, but was replaced by Obadiah Brown with a new coach house in 1801, after his father, Moses, had purchased the house.

The Dorr House, a brisk walk down Benefit Street to the corner of Meeting Street, is an equally formal estate, featuring a gleaming clapboard mansion with matching and connected outbuildings and carriage house. The work of Providence was that of architect John Holden Greene, and an article in the Rhode Island Historical Society Bulletin of April 1957 declared that "the placing of the house on the lot and the arrangement of the various appendage buildings has always been considered one of Greene's outstanding achievements in site planning."

Evidence uncovered at that time, with the discovery of Greene's plans, seemed to indicate that the shed, stable and carriage house were all built before the house in 1810.

The days of small farms and hogs and cattle freely roaming the streets of Providence were gone. Now horses' hooves clattered on the cobblestoned lanes and drew small, elegant carriages on the hillside streets above Market Square, where lines of wagons hauled by draught horses to the teeming docks could be seen along the river each morning.

To a large extent, it was horses that spurred the early economy, long before the highways and stagecoaches and the Industrial Revolution, when the proud animals whose ancestors had plowed our early fields were saddled with a life of drudgery and back-breaking labor. It was horses that transformed the economy of the colony in its early years, from the villages to the rural farms, where their story in Rhode Island began.

2

Horse Farms and Stables

The Narragansett Pacer and Horse Racing in the Ocean State

The import of horses to the shores of Rhode Island and other New England colonies began sometime after the settlement of Plymouth in 1620. We know from records that sixty horses were sent from England to Massachusetts in 1630 alone, though it is difficult to know when the first horses set foot on Rhode Island soil.

The animal was so scarce in the early colonies that when two travelers needed to get somewhere and they only had one horse, they used the "walk and tie" to get to their destination. One man started riding the horse while the other walked. After a set distance, the rider dismounted and tied the horse to a tree or post beside the road. The man walking would untie the horse when he reached it and ride ahead, passing the first rider until he reached the next point, and so on.

Within twenty years of the first horse reaching New England, the animal would become the chief mode of work and transportation. With such demand, here and abroad, the Rhode Island farmers were early established in the trade of horses.

William Coddington shipped a number of horses to the West Indies in 1656, though it was claimed by William Brenton that these were his Massachusetts registered horses, a herd he kept on the neck of Point Judith. A letter written in 1666 by Peleg Sanford would seem to indicate that by this time, the Narragansett planters were shipping horses on a regular basis:

In respect of the extremity of winter I could not gett your horses from Narragansett, but intend if your pleasure soe be to send for them as Soon as the ketch arrives which we dayly Expect.

A descendant of the Narragansett planters wrote:

Horses were the most important of the domestic animals, since they not only worked the farm, but were the means of locomotion. There were few roads in the South County in the middle of the last century [1700s], *but bridle paths led from one great estate to another, through endless gates.*[39]

Of all the commercial ventures of the planters, the one that became the most lucrative was the selling of horses, both draught and "riding beasts," as well as the breeding of horses, most notably, the development of the Narragansett Pacer.

John Hull, who was raising horses in the Narragansett Country as early as 1672 and had a farm in Sandwich, Massachusetts, was the first to consider developing a pure strain of horse to offer the market. He wrote to Benedict Arnold, who served as the secretary for Hull and the other partners in the Pettaquamsett Purchase, with his proposal:

Sir, I have sometimes thought if we the partners of Pointe Juda Necke did fence with a good stone wall at the north end thereof that noe kind of of horses nor cattle might get thereon…& procure a very good breed of large and fair mares and stallions and that no mongrel breed might come amonge them…wee might have a very choice breed for coach horses some for the saddle some for the draught…& in a few years might draw of considerable numbers & ship them for Barbados Nevis or such ports of the Indies.[40]

Inventories of a few of the planters' estates show that James Wilson owned "31 horse kind" in 1705–06, Rowland Robinson owned 64 horses, mares and colts in 1716 and William Gardner owned 30 horses and mares and 1 "young stone horse." Jeffrey Hazard registered his "riding beast" as well as "13 breeding mares, 3 geldings, 5 three-year-old mares, and a sorrel stone horse," which he valued at £400.[41]

Rhode Island ports were the first in New England to enter the horse trade, mentioned as early as 1681 by then governor Sanford as one of "the principall matters of export" from the colony. As the trade increased over the next twenty years, "Rhode Island vessels were taking horses to Jamaica, Barbados, Nevis, Antigua, St. Christopher, Monserrat and Surinam."[42]

By 1700, the development of the Thoroughbred had become well established in Rhode Island, Massachusetts and Connecticut, as well as in some southern colonies. The purebred horses shipped to England elevated equine sports to great popularity, and the breed was also used to improve the strain of Morgans and quarter horses used for stagecoach and wagon labor.

Advertisement from James Brown depicting a Narragansett Pacer. Hundreds of these horses were shipped to Surinam and to closer locales, such as Maryland and Virginia. *Courtesy of Brown University Archives.*

Draught horses were sold to owners of the sugar mills in Barbados, which still used horses to grind the sugar, as compared to the English windmills used in the same period. But just as John Hull had envisioned, the prized horses on the market were the purebred Pacers.

The consensus among most historians was that the horse that came to be known as the Narragansett Pacer was developed from a strain of Andalusian stallion and another breed. Miller, in his book *The Narragansett Planters*, speculates this might have been an Irish hobbyhorse, which shares many traits with the Pacer. A description written for the first American edition of the *Edinburgh Encyclopedia* in 1830 reads:

> *They have handsome foreheads, the head clean, the neck long, the arms and legs thin and taper, the hindquarters are narrow and the hocks a little crooked, which is here called sickle hocked, which turns the hind feet out a little, their color is generally, though not always bright sorrel, they are very spirited and hold both head and tail high. But what is most remarkable is that they amble with more speed than most horses trot, so that it is difficult to put some of them on a gallop. Notwithstanding this facility of ambling, where the ground requires it, as when the roads are rough and stony, they*

> *have a fine easy single footed trot. These circumstances, together with their being very sure-footed, render them the finest saddle horses in the world, they neither fatigue themselves or the rider.*[43]

The Reverend James MacSparran, who arrived in Rhode Island in 1721, would write that he had often ridden a Pacer "fifty, nay sixty miles in a day here in New England where roads are rough, stony, and uneven." Thomas Hazard would recall that "they had great endurance, and were capable of carrying heavy burdens in addition to their rider, and many a journey to Boston, or into Connecticut, did they make."[44]

An early English observer in Rhode Island included, among other critical views of American husbandry, that "a New Englander will ride his horse full sped twenty or thirty miles; tye him to a tree, while he does his business, then re-mount and gallop back again."[45]

So close is his account with others that it is almost certain that the writer witnessed the great population of Pacers that traversed the roadways. Such was the legend of the Narragansett Pacer, that similar stories abound. George Washington owned a pair of Pacers on his Virginia estate. One story has the future commander in chief riding a Narragansett Pacer in a local race in 1768. Another is that Paul Revere also favored the horse, and it was a Narragansett Pacer that carried him on his "midnight ride."

While these legends may be disputed by historians, what cannot be disputed is that the Narragansett Pacer was, for over a century, considered an invaluable breed for racing and riding.

As early as 1700, branding began to be practiced because of "the growing prevalence of horse thieves" throughout the colonies. As Deane Philips noted in his *Horse Raising in Colonial New England*:

> *The brander in most towns was a dignitary of no small importance, and as a rule was required not only to brand each animal but also to keep a record of the operation in an official book together with a description of the animal, and the name and residence of the owner.*

As demand grew for horses, regular buyers, or "horse coursers," surveyed the farms in the countryside, gathering horses to sell, though they were sometimes suspect and traded horseflesh by nefarious means. The planters and other large estates handled their own inventories and often shipped the horses from docks built right off the coastline of their farms, though they sometimes used the ports of Newport and Providence as well.

On February 3, 1735, merchant James Browne of Providence wrote to Samuel Cuttler:

> *Sr. These are to Acquaint you that I want some Horses that are in Case fitt for Shiping, that are worth between Seven & fourteen pounds, Mares will do if they are in good Case, they must be between three & Advantage & Eight years of Age, if you or any of your Neighbors have any such Naggs to Sell please to lett me see them between this & the 20th. of this Month.*[46]

Browne would continue to transport horses, writing again in February 1738 for "teen or a dusin Surnam horses." Hazard wrote that "this was a valuable animal, for which sugar, molasses, tea, and indigo were exchanged."[47]

Local trade occurred, but on a far lesser scale, as the price for horses in America fluctuated with the loose value of the currency printed by the colonies. By the mid-eighteenth century, Rhode Island was issued one bank after another, and the currency steadily depreciated as the credit of the colony declined. This is reflected in the purchases by Robert Hazard. A "three year old horse" cost £150 in 1750; a year later, he purchased "a thirteen year old bay mare with a white nose" for £55. In 1765, he paid £244 for "an old black trotting mare." Hazard's selling to locals seems to be just as rare, with an entry for selling "one old horse" to Rowland Robinson in 1767.

Still, for the planters, between 1700 and the outbreak of the American Revolutionary War, there was "a ready market for horses" in the West Indies, as well as along the southern coast of America. As early as 1739, the *South Carolina Gazette* advertised for sale "a Rhode Island Racing Stallion" in its pages.

Indeed, the Narragansett Pacer would become renowned for its racing prowess as well, a factor that kept horse racing from being banned in Rhode Island as it was in other northern colonies. Pacing races were often held on Little Neck Beach, now near Narragansett Pier, and the engraved silver tankards presented to winners became prized among the Narragansett families.

Thomas Hazard recalled in his memoir that an ancestor of his, J.P. Hazard, had told the historian Updike, who was then writing his history of Narragansett, that an elder neighbor of his, Enoch Lewis, had once told him that he'd been to Virginia as a "riding boy" or jockey for races and that such visits were common between the racing crowd of Narragansett and Virginia.

Hazard wrote that his ancestors were, like their English forefathers, "a horse-racing, fox-hunting, feasting generation."

The enthusiasm for horse racing was not limited to the Narragansett families. In Providence, shortly after the War of 1812, Edward Babcock purchased a large farm on the south side of the city, extending from Broad Street to the waterfront. His father, John Babcock, helped to finance a triangular trotting track on the farm to race his Pacers.

The races were so popular that around 1851, the track was redone and fenced in. By 1861, fellow enthusiast Amasa Sprague had formed a partnership with Babcock to operate a professional racetrack on a leased portion of the farm.

The "Washington Park Trotting Association" became an overnight success and hosted the Grand National Circuit for several years. When Babcock and Sprague reputedly bickered over gambling at the park, Sprague left and built his own track in the town of Cranston, called Narragansett Park, which soon outdrew the old track in Providence. The park became highly popular and hosted a race that included in its gates the legendary Dan Patch, a race horse as famous as Seabiscuit became in the new park decades later. The name would be revived in 1933, with the building of Narragansett Race Track in Pawtucket. A world-class track was laid out before spectator stands and a great complex that held twenty-two barns with stables for over one thousand horses.

The new Narragansett Park became even more popular than its predecessor, as the country had taken to Thoroughbred racing. As mentioned, the Narragansett Park hosted the famous Seabiscuit—seven times.[18] The racetrack remained a popular venue into the 1950s, when a slow decline began. The buildings and complex came to be in such disrepair that a fire begun in a hay barn in 1976 quickly spread to a pair of adjacent stables, killing thirty-six Thoroughbreds. The tragedy was the death knell for racing at the park, and it was closed by authorities within months of the fire.

Though, as with the park, the stables that were built to house these valuable horses are largely gone, what remains and the later stables that were built in emulation of those on the planters' estates give us a glimpse into the lives of those who not only owned and rode the Pacers but also those who worked in the stables and cared for the horses.

Some of the first of these would have been built in Newport. Long regarded and written of by visitor Alexander Hamilton as "being an entire garden of farms," Newport was also becoming a resort for those

Livery stable in Narragansett at the end of the nineteenth century. Horse and wagon took tourists from the docks to the great resort hotels. *Taken from* Narragansett by-the-Sea.

who had visited during the long years of coastal trade and became beguiled by the island.

Wealthy southern families were among the first to make extended visits to the city by the sea during the long summers. While living in luxurious homes down south, they came to revel in Newport's simplicity and unspoiled beauty. They took long walks on the beach, played lawn bowling, took sponge cake and fresh milk with afternoon tea and danced in the evening. There was an active social calendar printed and reported on in the *Newport Mercury*. By 1774, the city was the most popular resort in the colonies. This familial society was broken apart by the occupation of the city by British troops and Hessian mercenaries at the outbreak of the Revolutionary War. The British army routinely raided nearby farms for livestock and cut down magnificent trees for firewood.[49] They stripped the fine colonial houses of their belongings, even taking away the signature front stoops from the American houses so that inebriated British and German soldiers would not be injured tripping over them.[50]

The war, and its subsequent economic reverberations in the aftermath, kept Newport in poverty for some years. The British had razed many of the finer houses in town, and others had fallen into disrepair. Farms that had been abandoned during the war slowly came back to life but were mostly self-sustaining.

Around 1825, those southern families who had summered in Newport during the prosperous prewar years began drifting back, and soon, they were joined by others from Boston, New York and Philadelphia. Some wealthy plantation owners from Cuba also picked Newport as their resort of choice. As many of the fine houses were gone, however, the guests stayed in rooming houses or in small, family-run hotels that were rapidly being constructed.

As the century progressed, more tourists came to Newport, and grand hotels began to be built to accommodate the mass of vacationers, the first being the "Ocean House" built in 1844. Such was the need that when the hotel burned down after only a year of existence, the investors quickly rebuilt it, at nearly three times the cost of the original structure. While investors continued to build large hotels, Newport-born Alfred Smith took a different direction. A simple tailor who had found success in New York, Smith returned to Newport with a considerable nest egg and a plan.

Stone stable in Newport, Rhode Island. *Courtesy of the author.*

He began buying property in underdeveloped areas of town, and knowing the landscape by heart and where the most advantageous views were benefited his vision. The purchased property was then divided into lots, which Smith landscaped to make more attractive. In 1851, he partnered with Joseph Bailey and purchased a large tract of land that was largely seagrass and sand, with a lonely beach on the southern side of the island. He persuaded the town to extend Bellvue Street to his beachfront property and lined the newly named Bellvue Avenue with trees to resemble the grand avenues of Paris.

It was here that the southern families and other early visitors built their homes and the stables for their horses. In 1852, a wealthy southern planter named Daniel Parrish built a Florentine-style palazzo, which he christened "Beechwood." Two years later, William Wetmore, who had made his fortune in the China trade, built another Italian-style villa, which he named "Chateau-sur-Mer." The great house was built of granite at the site of a house previously constructed by James Allen, which had burned down before it was even occupied. The gatehouse and the stables, which were also completed by Allen, remain and are of brownstone to match the original house on the site.

The stables were initially constructed as a two-bay barn, whose exterior was board and batten and brownstone. When the chateau was expanded by architect Richard Morris Hunt in the 1870s, the stables were also improved, with Hunt hiring Newport architect George Champlin Mason to expand the stables by eighty-four linear feet, which allowed fourteen more stalls to be added. Over the next thirty years, the Wetmores added further additions, including feed-storage rooms and an exercise yard paved in concentric circles of Belgian block.

No expense was spared, and the stables, by the end of the nineteenth century, included all the "modern" amenities with elegant finishing.

A rooftop windmill powered a ventilation system that funneled hot air away from the hay loft, thus greatly reducing the risk of fire. The roof's drainage system included an attic cistern that channeled rainwater into the horse troughs. Attic bins also allowed feed to be dropped directly into troughs. The rooms were lined in red and yellow brick, with finishing touches made in beaded board and minton tiles in celadon or terra-cotta red quatrefoil patterns. Cast-iron lions' heads, imported from an Irish foundry, ornamented the controls for the hopper-style clerestories.

In these last decades, famous architects such as Charles McKim and Stanford White, as well as Hunt, were creating opulent "cottages," stables

The Wetmore stables in Newport, now part of Salve Regina University. *Courtesy of the author.*

and carriage houses for their wealthy clientele. One of the most extravagant and unusual houses was built for the banking heir Oliver Hazard Perry Belmont. The third son of August Belmont, the heir had horse breeding in his blood. His father had been an avid Thoroughbred racing fan and raised horses on his 1,100 acre farm in Babylon, New York. The prestigious Belmont Stakes, inaugurated in 1867, was named in his honor. Belmont was to build one of the first "summer cottages" in Newport with the construction of "By-the-Sea" in 1860.

Oliver's older brother, August Belmont II, raised polo ponies on the farm in New York, and when his father purchased a farm outside Lexington, Kentucky, Belmont junior administered the breeding business, which was to produce 129 American Stakes winners, including the famous "Man o' War."

The exercise yard and extended stables. *Courtesy of the author.*

Belmont Jr. organized the Westchester Racing Association and was one of nine founding members of the National Steeplechase Association in 1895. In 1905, he built Belmont Park and transferred the Stakes to what remains one of the most famous Thoroughbred racing venues in America.

Belmont also had horses competing in races in England and France and even established a breeding farm in Upper Normandy. His younger brother, however, chose to keep his horses, literally, much closer to home.

When August Belmont died in 1890, Oliver and his brother inherited a massive fortune. Having survived a brief scandal and a failed marriage, after which he nursed his wounds in France, Oliver returned to Newport, determined to build his own summer cottage. He hired Richard Morris Hunt to design "Belcourt Castle" with one distinction: the first floor of the entire mansion was to be occupied by stables for his beloved horses.

Author Cleveland Amory wrote in his book *The Last Resort*:

> *The Belmont horses had a change of equipment morning, afternoon, and evening. For the night they were bedded down of pure white linen sheets with*

the Belmont crest emblazoned on them. A barracks for a battery of grooms also occupied the first floor. Above the stables, in the salon of Belcourt Castle, Belmont kept two stuffed horses, old favorites of his, which were mounted by stuffed riders in chain armor.

The neighboring Vanderbilts were also highly competitive in equestrian competitions. When Cornelius Vanderbilt II constructed the Breakers, his luxurious marble mansion that replaced the original shingled cottage, the stables he built were no less opulent. In the era of high society, which has become known as Newport's Gilded Age, the culmination of the summer season was the prestigious horse show in early September.

A two-story, U-shaped structure, measuring 100 feet deep and 150 feet wide, the stables held twenty-six tie stalls and two box stalls, with a large hayloft and grain room above. The rooms for the twelve stable boys and grooms employed were also on the second floor of the north side, while in the rear of the building on the south end, there was a large kitchen, dining room and living room for their use. The head coachman, who supervised the stable hands, held a five-room apartment in the building. Unfortunately, the second floor of the structure was destroyed by fire in 1970. The Newport Preservation Society's description of the Breakers

Old postcard of "Belcourt." *From the author's collection.*

gives us a glimpse into the life of the stables during the years that Alfred Gwynne Vanderbilt occupied it:

> *Mrs. Vanderbilt sent down a day-book every morning at eight o'clock with a list of the carriages that would be used that day, and people in the house would call down their requests on the telephone. When a carriage was requested, the horses were brought out, hitched, and left the building from the north door. All the returning carriages entered through the south door. The carriages were unhitched, washed off, and wheeled into the carriage house. The horses were taken to the back, unharnessed, and washed down in two rooms with cement floors.*

Vanderbilt was an entrepreneur in the sport of coaching and had the Brewster Company of New York—the finest coach builder of the times—design and construct the Venture, a luxury coach that he brought to England each summer, along with the stable boys and grooms to run a stagecoach line from Brighton to London and back again.

The love of horses was to have a long lineage in the family. After Alfred Gwynne Vanderbilt's untimely death aboard the *Lusitania* in 1922, his son William Henry Vanderbilt III inherited Oakland Farm in Portsmouth. He settled on the farm with his new wife, Emily O'Neill Davies, and continued the stud farm his father had established there. He was to have an active political career, serving in the Rhode Island Senate from 1929 to 1935, and as governor of the state from 1939 to 1941. He would also continue his father's coaching business, using the Venture and its team of six horses to ferry visitors from New York to Newport and back.

He called this coach service "the Short Line" and later began a bus coach company, which he named the same. Vanderbilt loved to bring the stagecoach out of retirement to horse shows throughout southern New England well into the 1930s.

In the town of Portsmouth by the late 1800s, the old families who had founded the settlement were fading from the landscape, and parcels of colonial farms were sold off by the descendants.[51]

Industrialist A.C. Taylor purchased several adjacent properties and established a "gentleman's farm" where he could commit more time to his passion of breeding horses and equestrian sport. The estate was named Glen Farm and was as grand as any of the Newport properties, with a French chateau designed by John Russell Pope and featuring stone- and wood-framed barns and stables of massive size built as the farm expanded, with

Left: An invitation from William H. Vanderbilt III and his wife to a "barn dance" in 1936 at their farm. *From the author's collection.*

Below: One of the great stone-and-timber barns at Glen Farm. *Courtesy of the author.*

the year of construction neatly placed in the front cornice of the two-story rough stone structures. In addition to the stables, there was a mill, a wagon shed, blacksmith shop, tool shed, icehouse, pottery shed, animal hospital and pump house.

Taylor's farm became nationally known for the blue-ribbon Clydesdales and Percherons it produced for the prestigious horse shows. The farm passed to Taylor's son in 1921 and through later generations until 1982. The last of the Taylor lineage, Mason Phelps, established the farm as host to the International Jumping Derby until his retirement to Florida. The Phelps Trust then sold the farm to the town for $3.6 million.

In time, the town of Portsmouth found the upkeep of the farm prohibitive, and it fell into disrepair to the point of nearly being slated for demolition. Boston investor Dan Keating rescued the farm, signing a ten-year lease with the town, during which time he spent millions of dollars in renovations and revitalizing the grounds with polo matches on a regular basis during the season.

With a respect toward the history of the farm, Keating paid close attention to detail during renovations, even to the point of commissioning the brass hardware to be recast to match the originals.

When entrepreneur and philanthropist Samuel P. Colt purchased his four-hundred-acre waterfront property called North Farm on Poppasquash Point in Bristol, he was determined to live the life of a gentleman farmer, and his estate and stables decidedly emulate the Newport mansions.

A pair of bronze bulls adorn the marble balustrades at the entrance to a long, linden-lined drive that led to the estate. A large, stone barn and stable with twin wide, round silos, a gabled roof and an arched entryway into the barnyard was built on the grounds, with riding paths running beside the waterfront and along the border beside the far humbler Coggeshall Farm.

Of course, outside these grand estates and "plantations," there were many farms of more modest means whose stables reflected a simple, square-framed design but whose flourishes—such as a cupola, decorative trimming, pigeonholes and even swift houses—reflected either the builder's ethnic tradition or, even likelier in America, an owner who could afford to indulge his own particular taste.

The majority of these "one-horse stables" were built in more urban areas, not only due to less building space available but also because an urban dweller by the mid-nineteenth century would be far less likely to have any other livestock than say, a person living just a few miles out of the city. Many of these can still be seen today in the older neighborhoods

Entrance to the stables on Glen Farm. *Courtesy of the author.*

The stone barn on Colt's Estate, now Colt State Park. *Courtesy of the author.*

Stable of Kingscote Estate in Newport, Rhode Island. *Courtesy of the author.*

of Rhode Island's cities and towns. Those still standing were converted in the early twentieth century to garages for the family automobile. In some cases, they still hold the features of their earlier use: a gambrel roof; a cupola that once vented the hayloft; or the window above the garage door, where the swinging doors of the hayloft once hung, to take in bundles by pulley from a wagon parked below.

Today, the horse is again primarily a rider's companion and respected for the beauty and grace the animal displays in riding, racing and equestrian

A common stable transformed into a garage. *Courtesy of the author.*

competitions. While horse racing in the region has declined, the popularity of equestrian competition has grown in Rhode Island, an event that historian Walter Nebiker, writing of the sudden increase of horse farms in Burriville in the 1980s, called a "post-war phenomenon."

In Rhode Island today, over fifty horse farms board, feed and exercise animals, many of whose owners' properties might no longer accommodate stables. Owners train with their horses regularly and often take them on the weekend in trailers to riding groups or competitions. Often, these modern stables come equipped with a riding ring for practice and even a jumping course.

Much like the Ballard Farm, the old Winsor Farm in North Smithfield has evolved from its one-time use as a dairy and produce farm to become a first-rate riding academy. A new, modern barn houses the horses used for training in the paddocks and riding courses on the property. Bill and Abby Lowry also provide CEM quarantine services, as well as materials to construct a home equestrian course, as well as selling horses raised in the stables.

Other long-established horse farms, such as historic Paine Farm in Foster, still breed horses as they have done for generations, as well as provide riding trails, horse-drawn cart rides in summer and sleigh rides in winter on their expansive properties. Paine Farm dates back to 1785 and still retains 110 acres and an apple orchard. The farmhouse on the property dates from 1835.

Many families who own horses today modify their property to suit their needs or lease an already existing farm, as do Randy and Donna Levesque of Tiverton. The farm where they and their horses are living was originally the Stephen Cook Farm, built in the mid-nineteenth century, and it still retains the original house and wood and stone barn. There is an additional wooden-shingled double cupola barn behind the house, converted into stables some years ago by the current owner under the name of MMM Horseman's farm. The Levesques house five horses in the barn, a few of which Randy trains and rides in equestrian competitions. He has built his own practice course and exercise ring in the pasture behind the barn.

Other farms are used as venues for competitions; in fact, the old Glen Farm is the site of many of these in season, as well as remaining a venue for the International Polo matches that are played each year from spring into fall.

3

Tavern Barns and Stables on the Stagecoach and Trolley Routes

As roads improved in the eighteenth century and towns grew in people and traffic, the necessity of taverns, or "ordinaries along public roads for meals, drink, and lodging," caused county authorities to license many houses on the stagecoach routes of Rhode Island. In this age of transportation, the meals, drink and lodging applied equally to the rider's companion. Taverns who sheltered and had feed and supplies for horses were to become well known and visited frequently. Those with less-cared-for stables were likely to be occasioned only by locals or an ill-informed traveler, as many letter writers expressed with displeasure to friends and family back home. Traveling from Providence to Philadelphia in a chaise loaned from a friend, young William Rogers recorded that some hours after leaving Providence

> *we all din'd at little rest one Potters Tavern-fard' but poorly—no oats or hay for our horses—Roads for some part of ye Way past Description Bad.*[52]

The first developed and maintained route in Rhode Island was the Post Road, beginning at the northern boundary of the state in Pawtucket and winding southward along the coast through Providence, Cranston, Warwick, East Greenwich and farther into Kingston, where Potter's Tavern was located, and then to Westerly. The first taverns in the colony to accommodate travelers and their horses and later carriages and wagons populated this route, which was part of the larger Post Road leading from Boston to New York.

Among the first would certainly have been Steven Jackson's large stone-ender, built at the intersection of the Post Road and Connecticut highway, two miles from Providence. The stone-ender would be expanded to twice the size by 1700, with a English-style barn just north of the house and a larger barn for horses and carriages erected across Pidge Lane by the 1800s when the Sayles and Pidge family owned the tavern.

As for Providence, an old penciled drawing dated from 1775 shows a large hay barn on Sabin Street, just above the tavern where James Sabin, for whose tavern the street was named, "kept a place for man and beast." The Hoyle Tavern, at the site where Obadiah Brown established a tavern in 1739, was a bustling hotel by the mid-nineteenth century, with a barn as big as the house behind for horses and carriages. The whitewashed barn featured a central cupola, adorned with a cast-iron weathervane, which depicted a horse and rider in a four-wheel sulky.

A long-standing livery was kept on Orange Street opposite the Opera House, which proved to be its equal in longevity. By 1900, L.A. Tillinghast's restaurant was advertising carriages "for all who wish to be taken to our restaurant after the theatre WITHOUT CHARGE."

On the route south of town, the traveler would have noticed the White Swan Tavern in the small community of Burgess Cove. A small brewery

An old postcard of a stagecoach and riders on the Centerdale–Harmony Route, circa 1880. *From the author's collection.*

Advertisement, circa 1900. *From the author's collection.*

had been established there, but these were only ruins by the mid-nineteenth century. The tavern was long a favorite watering place for the privileged sons of Providence merchants, and it was just far enough from the tony streets of the East Side to give them license to carouse.

The village of Pawtuxet, a five-mile ride from Providence along the Post Road, offered several taverns for repast, with barns for shelter if the weather was bad. If a repast was all that was required, the gentlemen of Providence might go for "an outing" to "the Willows," a "gentlemen only" club that featured prominently on the route in Old Warwick. Built around 1850, the club had a large barn, as big as the house on the property, which sat adjacent to a small pond where "clambakes" were often held.

A popular tavern in the town of East Greenwich on the northeast corner of Main and King Streets was Tibbets Tavern. For many years, a post stood on the corner where the tavern sign was hung, adorned with the coat of arms of the state of Rhode Island. A large ash tree grew in front of the building that shaded the whole of the front yard. The tavern was

> *a large, long, white house with small-paned windows and wooden shutters. The door in the center of the building was graced by a handsome fan light. Four or five double steps ornamented by an iron railing led up to the front door. A side door, on the north end, led to the part of the building used for retail stores.*[53]

A large barn where horses and carriages were stabled stood on the east side of the tavern. John Tibbets owned the tavern as early as 1820, and it was operated by his colorful wife, Susannah, until about 1853, when his nephew and namesake took over the tavern. During the ten years that the younger John Tibbets ran the tavern, a Masonic Hall was added on the second-floor level, with a covered stairway on the northwest side of the building.

This addition was to be the site of many dances, banquets and, for many years, Thanksgiving dinners held by the Baptist Church. When John Tibbets moved away, the tavern and stable were run by James Fones, a man who had previously held a small homemade cigar shop.

One summer night in 1872, the barn was struck by lightning. Fones ran down Main Street, alerting the neighbors, and soon, a fire-bucket brigade was formed from the hand pump in front of the courthouse. By the time this was organized, however, the fire had spread to the tavern. Men quickly sat astride the ridgepoles of nearby houses to douse their roofs with water as the old landmark and its outbuildings continued to burn. The tavern, most of its belongings and a dozen or more horses in the barn were lost in the fire.[54]

As the Post Road wound into Kingston, Charlestown and Westerly before the Connecticut line, other stagecoach stops and taverns flourished at the height of travel by horse and coach. Haven's Tavern received early mention in the journal of Sara Kemble Knight, an accomplished horse rider, frequent traveler and, thus, a severe critic of the roads, taverns and people whom she met in her three-day journey through Rhode Island in 1704. She found taverns filled with locals who were loud, drunk and often engaged in arguments with other patrons about one thing or another. Recent rains at the time of her travel had swollen local rivers to the extent that crossing them became a harrowing ordeal:

> *About Three afternoon went on with my Third Guide, who Rode very hard; and having crossed Providence Ferry, we come to a River wch they Generally Ride thro'. But I dare not venture; so the Post got a Ladd and Cannoo to carry me to tother side, and hee rid thro' and Led my hors. The Cannoo was very small and shallow, so that when we were in she seem'd redy to take in water, which greatly terrified mee…*

And this was only the first in several rivers she would cross that day. Ms. Knight managed to guide her horse through another as the route headed into southern Rhode Island, but

> *here We found great difficulty in Travailing, the way being very narrow, and on each side the Trees and bushes gave us very unpleasent welcomes wth*

> *their Branches and bow's, wch wee could not avoid, it being so exceeding dark. My Guide, as before so now, putt on harder than I, wth my weary bones, could follow; so left mee and the way beehind him.*

Eventually, as they reached the town, the road became easy, and when at last the guide sounded his horn to inform her that he had arrived at the coach stop, she recorded in her diary: "That musick then was the most musickall and agreeable to me."

By 1730, the Post Road had been widened and greatly improved into South County. As carriage traffic increased, taverns and barns were expanded, or new ones were built to accommodate people and the coach horses. When Haven's Tavern was severely damaged by fire, another tavern was built on the long-popular site by William Maxwell in 1803. It is said that some of the original hand-hewn timbers from the old tavern were incorporated into the house. When Maxwell died, he deeded the property to Henry Warde Green of East Greenwich, and the property included 128 acres and the house, as well as the "barn, crib, and other buildings."

The large barn on the property is a one-and-a-half-story, one-by-two-bay structure with clapboard siding and a long short-paneled door. It rests on a rubble stone foundation with a gabled roof.

The house and outbuildings remain as property of the iconic Pagoda Inn, established in 1949 and still operating today.

Farther along the Post Road, the Stanton Farm became an inn around 1810, and the stables that once held forty horses for breeding were used to house the teams of coaches and the occasional horse for the individual rider. The Stanton Inn was to serve as a tavern for nearly two hundred years, falling into disrepute by the late nineteenth century and known as a gambling house in the 1920s. It was restored later in the century and run successfully as an inn and motel.

Nearby, Sidney Gavitt built a large house close to the road on the family dairy farm, whose land had been deeded from the sachem Ninigret in 1731. The Ocean View, as Gavitt called his boardinghouse, became well known as a family establishment and remained popular for many years. The Ocean View also housed a small post office. Guests were "driven" to and from Westerly by neighbor "Billy" Latham, who owned a large stable nearby. A recollection written by Carol Burdick for a local newspaper in 1948 describes a typical summer at the resort:

The Gavitt Boardinghouse in an undated photograph. *Courtesy of Patrick Verdier.*

> *Each sunny day a group of mirthful boarders would drive in old carry-alls across the fields and through the farm gates down to Quonconontaug Pond. Leaving the horses there, Captain Gavitt's sons would sail them across, and they would bathe and picnic on what is now called Shelter Harbor Beach.*[55]

In an old, undated photograph, the great house looks like it is built on little more than a cart path, and the large, whitewashed barn is similar to the 1810 barn on the Ballard Farm, so we may assume it was built at the same time as the boardinghouse. When the Gavitt House was closed briefly near the turn of the twentieth century, it was purchased by one of its long-term patrons, Thomas W. Wiles. His daughter, in turn, sold the house in 1924 to Ward Sherman, and it became the Haversham Inn.

As other long-standing "roads" were improved from country lanes into main highways and stagecoach routes during the early part of the nineteenth century, more ordinaries were opened to travelers. In the rural areas of Rhode Island, these were often little more than simple farmhouses and barns. The Elihu Fish Tavern in Scituate, Rhode Island, is one example. The tavern was built by James Thornton in 1759 as a one-and-a-half-story structure used as his residence. Several additions were made to the western side of the house, and then in 1793, new owner Theophiles Blackmar "added a full half-house to the east side consisting of a second floor ballroom, first-floor dining room, and a cellar kitchen, all necessary elements for a tavern on the Great North Road from Providence to Plainfield CT."[56]

The tavern got its name from the man who purchased the tavern in 1801. Mr. Fish ran the tavern for twenty-five years. The English-style barn, perhaps one hundred yards down the road from the tavern, likely dates from this same period.

Many taverns in the more rural areas were constructed at intersections of major stagecoach routes. The farmhouse and connecting outbuildings, including a large barn at the end that stretches out along a curved drive off Douglas Pike near the intersection of Pound Hill Road in North Smithfield, is but one example. Originally the Nichols Hotel, the Federal-style farmhouse and carriage house were originally set well back from the road behind a spacious lawn and enclosed with an elegant cast-iron gate, with a formal walkway to the house. Early drawings of the property reveal

The barn of the Elihu Fish Tavern in Scituate, Rhode Island. *Courtesy of the author.*

The Old Nichol's house and barn at the intersection of Douglas Pike, Mowry Road and Pound Hill Road. *Courtesy of the author.*

that this dwelling was at one time a much more formal establishment than the saloon-like Western Hotel that was built farther down the pike in 1805.

By the peak of stagecoach travel in the early to mid-nineteenth century, those areas around the intersection of well-traveled routes would hold a community of taverns, inns and hotels. In early photographs hung on the wall of the historic Stagecoach House Inn at the junction of Nooseneck Hill Road and Route 138 in Wyoming, Rhode Island, the intersection resembles a bustling western town—a reminder that not all of New England was white picket fences and church spires.

Along the coastline, towns that had transformed themselves from farming communities to travel destinations with hotels, restaurants and long, sandy beaches that accommodated nineteenth-century bathers used horse and carriage, as well as wagons, for a different labor.

Tourists in those days did not travel light, so along the docks where these travelers arrived, liveries were built for the horse teams and wagons that carried the tourists and their baggage to the hotels when they arrived and back when it was time to depart. For the day traveler, horses were rented by the hour or for the day. Some liveries even ran routes of their own in fancily decorated carriages, taking tourists along the scenic routes of Narragansett Bay.

Early photograph of the stagecoach house at Wood River, circa 1869. *Courtesy of the author.*

In the urban areas, with the advent of horse-drawn trolleys for public transportation, massive barns were built to house both horses and trolleys. The first organized trolley service in Rhode Island was established in 1864 and traveled from Central Falls, through Pawtucket on the Post Road, into Providence and then back again. Six to eight quarter horses pulled the trolleys along the route, getting passengers to their destinations within fifteen minutes of travel each way. Amasa and William Sprague created the Union Railroad in 1865, in large part to ensure that workers arrived on time for shifts at the brothers' Cranston Print Works. By 1873, when the Spragues' fortunes declined, the horse trolley service was purchased by a group of stockholders, including Jesse Metcalf, part owner of the *Providence Journal*. Metcalf was instrumental in the transformation from horse-drawn to electric trolleys, the first of which traversed the streets of the city in 1892. Marsden Perry soon had a monopoly on the Providence trolleys and enjoyed that privilege until the advent of the automobile.

Copeland's Livery Stable on Benefit Street, just across the lane from the Old State House, serviced coaches and riders since at least the middle of the nineteenth century. The stable was converted to the What Cheer Garage and serviced horses, trolleys and automobiles in the early part of the twentieth century.[57]

What Cheer Garage in Providence, Rhode Island. *Courtesy of the author.*

The terra-cotta plaque on the wall of the Summer Street Stables, Pawtucket, Rhode Island. Photograph by Rachael Keough.

Similarly, for some time before James Pidge's death in 1901, his large "horse and carriage" barn was used by coaches and then the trolleys. In 1905, James Ballou, who had purchased the Pidge Tavern from the family, sold the barn and tavern outright to the "railroad company," which was then preparing to widen the road to accommodate the trolleys with the still present horse and pedestrian traffic.

In downtown Pawtucket, the well-known Federal Furniture building was once the Summer Street Stables for the Post Road trolleys. The large, brick structure held stalls for more than 150 horses on the ground floor with carriage parking above and hay stored on the third floor. The stable also held a blacksmith shop and offices, as well as a waiting room for those taking the trolley to Providence and beyond. With its long-faded noir advertisement painted on the wall facing Main Street, the building only bears witness to its earlier use in the shadowed lane off Main, where the wide stable doors and windows are revealed and the motto is etched on a terra-cotta plaque above the doors: "How Do the Beasts Groan!"

A testament from the days when horses did so much labor for us, and while the electric trolley was hailed as "so much faster and gentler" than the horse-drawn carriages and trolleys that serviced the cities for so long, the automobile manufacturers first measured the strength of their engines in horse power, a term that still remains in use today.

4

Dairy Farms and Barns, Cattle Breeding and Rhode Island Cheese

The first known large dairy herd in the colony belonged to Richard Smith Jr. on his farm at Cocumscussoc, near Wickford, where the family had established their trading post. The farm held thousands of acres of grazing land, and the large herd of cows that roamed the shore fed on salt hay, which gave the cheese that Richard's wife, Esther, produced from the milk a distinctive flavor. The cheese became a well-known staple of the farm, and at the time of Smith's death in 1692, his 135 head of cattle were producing fifteen thousand pounds of cheese per year.[58]

The neighboring Narragansett planters also exported cheese from their large estates. The inventory of James Wilson in 1705 shows that he owned 72 head of cattle. Rowland Robinson also made "a large dairy," and according to one account, "his fancy was to have none but what were called 'blanket cows,' that is, cows that were entirely white all around the body between the shoulders and hips." Robinson attempted to keep "exactly one hundred 'blanket cows,'...neither more, nor less," but he struggled, even with raising and purchasing animals, to keep the herd at that number. Robert Hazard, who owned the largest property, is reputed to have kept 150 cows. His descendant, Thomas R. "Shepherd Tom" Hazard, wrote in his memoir that his "Father's Grandfather" held "large cattle ranges lying adjacent to Worden's Pond, several thousand acres, two thousand of which lay in the rich southern portion of Boston Neck, on the Tower Hill slope adjoining Governor Robinson's estate on the north."[59]

Hazard also wrote, in a curious anecdote, that he recalled a large painting once hung in the home of Colonel John Gardiner that depicted a wreck that had run aground on Westqueage Beach sometime in the previous century:

> *The ship was from some port in Europe, and was freighted in part with live cattle, which were graphically pictured struggling towards the shore amid the lofty breakers.*[60]

Caroline Hazard wrote that after the death of the widow Sarah Hazard in February 1772, her cows were taken away a week later and that "they were probably the wide-horned red-skinned cows of Devon extraction which have come to be the native of Rhode Island, giving milk rather scanty in butter-making properties, but excellent for the cheese which Narragansett became famous for."[61]

The word "dairy" comes from the middle English "dey," the word used to describe a female servant. The production of dairy products was almost wholly done by women, both in England and the colonies, where

Red Devon cows and calves today on the Watson Farm in Jamestown, Rhode Island. *Courtesy of Historic New England.*

many who were slaves and selected for their habits of cleanliness were trained for dairy work.

The first "dairy" on the farm was part of the house and was usually in a shaded corner with a recessed floor, or even the basement, if necessary. It was often called the "dairy chamber." But by the mid-seventeenth century in America, the "dairy" had moved outside. As with barns, the outbuildings built as "dairies," or "milk sheds," varied somewhat in design and decorative style, while serving the same function on the farm. The most common had a large, pyramidal roof with widely overhanging eaves above the square structure. Most held a single room, though some would have a second small chamber where dairy pails, crocks and utensils were stored.

These dairies mushroomed on the larger farms and estates, whose cattle provided much of the dairy products to an expanding community, and this alone may account for their growth. As historian Michael Olbert writes:

> *The quest for cleanliness created a class of small buildings fitted with brick or stone floors, their ceilings and interior walls plastered and whitewashed into a dazzling bulwark against the dirt. Because coolness was also vital, the floors of most dairies are two to three feet below grade and the dead space between the exterior wall and the lathed and plastered interior is packed with brick noggin or sawdust…instead of windows, dairies have long, horizontal openings high up on their walls, beneath the wide eaves. It's passive cooling: the high summer sun cannot penetrate the cool, dark interior of the structure.*[62]

The labor of producing dairy products was long and tedious, and any sudden change in weather—a frequent occurrence in Rhode Island—could spoil hours or days of work. Consider the process for making butter. After milking the cows, the dairy maids would strain the milk to remove the cartilage-like chunks, hair and flies that had drowned in the milk in the barn and then pour the milk into shallow pans, leaving it to sit for a day or two on shelves while the cream rose to the surface. The slotted openings were covered with gauze, or "cheesecloth," to keep out flies. The cream was then skimmed off the top with a large, flat spoon and stored in salt-glazed crocks. Cream has a longer shelf life than milk, as it is mostly fat. To make butter, the cream would be worked in a "plunge," or butter churn. When the clumps of butter were removed, they were worked together with a pair of wooden paddles called "scotch hands" to exude the last bit of water. Once this was done, the butter was rinsed with running water until the water ran

Above: The stone "milk-parlor" of the Chase Farm in Lincoln, Rhode Island. *Courtesy of the author*.

Below: Colonel Stanton's "Great Dairy" on the Old Post Road in Charleston. *Courtesy of the author.*

clear and then packed and salted in large crocks, where it could stored for up to two or three years.

In the making of cheese, the women would add rennet to the milk, which is taken from the lining of a calf's stomach. Its enzyme causes the milk to curdle and the process to begin. Within half a day, the milk and rennet would have formed a consistency of gelatin, and the "cheese" would be cut into squares. These solid curds would then be placed in a cheese press for several hours to allow the whey, or the watery residue, to weep out from the curds.

Soft cheese, or "farmer's cheese," was often made by wrapping the curds in cheesecloth and hanging them from a tree.

Of the planters, Robert Hazard is said to have employed twenty-four women to work in his dairy, producing twelve to twenty-four cheeses per day.[63] His son Thomas, who inherited the farm, maintained cheese as its most important product, producing 3,627 pounds in 1754 and 2,769 pounds the following year. As the price of cheese and dairy goods climbed, the farms produced less cheese but still made a considerable profit.

Others whose farms produced large quantities of cheese included Colonel Stanton, who "made a great dairy," and Rowland Robinson Jr., whose herd reputedly produced two pounds of cheese from each cow per day. The Sewell Farm on Point Judith Neck produced 13,000 pounds per

year. Nathaniel Hazard's dairy produced as much as 9,200 pounds in one year. Another producer of great "wheels," Hazard sold twenty-eight cheeses in 1756 weighing 2,836 pounds in total, or about 100 pounds per wheel.

The cheese produced on these farms in the eighteenth century was widely exported to the West Indies as well as to the neighboring colonies at home. It was sold in Benjamin Franklin's Philadelphia shop, where it was known simply as "Rhode Island cheese" and had an enviable reputation in Boston as well.

Sadly, no traces of the barns that held these large herds remain, nor do any of the outbuildings in which so much of the work on the farm was done. By the time of Thomas R. Hazard's *Recollection of Olden Times* in 1879, the eighty-two-year-old descendant of one of the great Rhode Island planters wrote:

> *A stranger now visiting Narragansett and observing the worn-out appearance of most of the farm-houses and lands, the latter to a great extent disfigured with dilapidated stone walls and loose boulders and cobblestones, and fast being overrun with briars and bushes, could hardly believe that scarcely a century ago this beautiful, though now desolate looking, farming country, teemed with superabundance of dairy and other agricultural products, and was studded throughout with princely mansions.*[64]

Though the era of the Narragansett planters had ended, dairy farms continued to grow in other communities of Rhode Island.

If there was said to be a remnant of the farms from the planters' time, one could be the Fry Farm. Long after its days as a "plantation," the farm remained predominantly a dairy, selling butter, cheese and meat in order to buy tea, coffee and spices. The original farmhouse burned in 1793, and the one built to replace it still stands today, though barns and outbuildings have been built several times over, being damaged or knocked down in storms, and replaced by the nine generations of the family that continued to maintain the farm.

Another long-standing farm lies on the land first improved by Benjamin Church in Little Compton, where in 1676, he built a house along the shore. As his farm and family grew, he built a second house in 1724, which remains on the property today. The farm came into the hands of William Wilber, and it was then sold to the Nelson family in 1756. They established a dairy they named Blue Flag Farms.

Ruins of the early Sheffield Barn in Little Compton. *Courtesy of the author.*

Initially, the Nelsons produced only small amounts of product to sell from the farm in Newport and Providence, but in time, milk production increased, and a windmill was also built to grind the grain grown on the farm into flour. A second house was built around 1830, and it still stands today. Two barns were originally on the property, but these were replaced by one large structure in 1885—the barn that is still in use today. A family story is that for many years, the milk was taken by buggy to Sakonnet Point to be shipped by steamboat to Providence.[65]

The farm that Israel Shaw inherited by marriage in the 1730s and became known as Winshaw Farm underwent its own evolution from a long period as a subsistence farm to a dairy after Frederick Shaw founded an ice cream business in Fall River in the late 1800s. Shaw would also resort to shipping his cream by boat from Little Compton.

In the early 1920s, the farm was inherited by the Wilder family, whose brothers Horace and Merrick established a stanchion-type dairy operation, transporting milk to the Fall River Dairy from 1939 until the 1980s. (As of 2008, Merrick was running the farm and its remaining thirty-five acres with the help of his son Emerson and grandchildren).

A nineteenth-century barn on the Samuel Hill Farm in Lincoln, Rhode Island. *Courtesy of the author.*

Samuel and his wife, Amey (Mowry) Clark, purchased farmland in the village of Albion in 1783. They built a house, barn and outbuildings to support the large farm. While their eldest son, Joseph, would leave the family farm for the sea, their younger son, Samuel Jr., married Rebecca Cushman and raised four children on the farm. The Clark-Church Farm passed through six generations of the family. The Church family named the farm "Kirkbrae," an expression in Irish meaning "church on a hill." The last member of the family to run the farm was Gilbert "Bertie" Church, who raised prized Ayrshire cattle on the farm and was a founding member of the Rhode Island Ayrshire Breeders Association. The Ayshire cow is larger than a Jersey and considered more docile, being undisturbed by noise or motion nearby. An old farmers' guide wrote of the Ayrshire that "she will give down her milk no matter what is going on around her."

A large, modern barn was built attached to the older, smaller structure that had sheltered earlier generations of the herd. A separate silo stood just feet beyond the main structure. Bertie Church ran Kirkbrae Farm until his death in 1949.

The story of many of these farms is illustrative of the evolution of the dairy from a small industry producing milk and other products for home use and local sale to farms that mass produce milk, butter and cheese sold at markets in distant locations. The advent of the railroad in Rhode Island proved to be a boon to local farmers. As with other communities, shortly after the first train station was built in Greene in 1854, farmers within three or four miles of the station began shipping their milk to Providence on the rail. As Squire G. Wood recalled, "A very early milk platform was built for the use of the milkmen, and a milk car added to the Plainfield train to Providence. The milk car ran continuously for sixty years or more, used by nearly all the farmers in the vicinity."[66]

With mass production, it became necessary to improve the quality of milk, and with improvements in farm safety standards and tuberculin testing for cattle, milk became a safer product. New technology, such as the Mehring Milking machine, which came into prominence in the 1890s, as well as commercial milk bottles, pasteurization and refrigeration, gave young farmers an opportunity to keep an arm's length from the sprawling city and industry that ran contrary to their nature but still be able to earn a living off the land in the twentieth century. A 1905 report from the state's Department of Agriculture optimistically proclaimed:

> *Many of the farms that were deserted are again being improved...the dairy farm industry has greatly increased during the last few years and our farmers are producing many hundred cans of milk that are being shipped to the cities to supply our city friends with milk and cream.*

One of the first of these modern dairy farms was established at Cocumscussoc by Austin Hoppin Fox in 1919. The Fox family transformed the long-neglected estate into a successful dairy concern. Cocumscussoc Dairy would eventually operate a milk route to Wickford, as well as an ice cream and milk bar at the corner of West Main Street and Post Road in town. The farm also established a purebred milk cow named the Cocumscussoc Ayrshire, which reputedly produced more milk than any other breed of its time. The herd of 145 was housed in a massive barn, and the complex also included a cooling plant, stock barn, tool shop, boiler room, a new dairy and a modern laboratory for testing the milk.

A similarly foresighted farmer was a young first-generation American named Walter Theinert, whose father had established a horse farm just up the road from the Clark Farm in Lincoln, Rhode Island. Both Walter and his brother William, are each listed on the 1910 census as "laborer on horse farm." By the age of twenty-five, however, Walter had become restless and was living at his brother-in-law's home in Kent County, working as a molder in the foundry where his in-law served as foreman. By 1925, he was back on the farm with the newly gained skill of masonry. In 1930, both he and his brother were listed in the census as "general farmers" on their seventy-year-old father's farm. But Walter had ambitions and ideas that would bring the farm's business fully into the new century.

By the 1930s, Theinert had built a massive stone barn in which he held his Guernsey cows, which produced that "special Golden Guernsey 'A' Milk" as advertised on his delivery trucks, some of the first that people in Albion had seen. Theinert's dairy delivered milk, vegetables, fresh fruit and chicken to residents in Lincoln and Cumberland, as well as the stores nearby and in Pawtucket and Providence. Walter ran the farm until 1942, when he sold the land and buildings to his brother William.

By 1948, much of the land was parceled for housing, and the farm was gone. The old stone barn remains today, as part of a private club, with tennis courts adjacent to the barn rather than a cobbled courtyard. Across the road, the Kirkbrae Farm has been the Kirkbrae Country Club for over thirty years, though the old farmhouse remains at the fringe of the groomed lawns and greens of the golf course and the parking lot, where cattle once grazed.

Numerous farms in the state would undergo similar transformations. The Nate Mowry Farm, where his famous inn stands, was sold to the Jordon family in 1841, when they established a dairy farm on the large swath of land. An early twentieth-century postcard shows the large cattle barn and silo that stood on the property. Farther southeast along Great Road, Benjamin Chase purchased the land Thomas Arnold had settled two hundred years earlier and established a dairy business in 1860, constructing a house and a massive complex of barns, which would eventually house over one hundred head of cattle. In 1925, lightning struck the barn, and the resulting fire destroyed much of the complex. A large stone garage, which was used as the original milking parlor, was unscathed. The owners then purchased the farm adjacent to their property, and the Chase-Butterfly Farm, as it was then called, continued to produce and deliver dairy products to the area until 1965. It was, in fact, the last of the great dairy farms in the town.

In neighboring communities, the evolution of farms was much the same. While not officially a part of the state until 1862, East Providence brought to the landscape a number of large farms, chief among them being White Rock Farm, which held four hundred acres of orchards, potato and corn fields and meadows. By the end of the century, Alfred J. Kent established a large dairy on the farm as well and became so

The Theinert Dairy barn as seen today. *Courtesy of the author.*

The massive barn complex of the Chase and Butterfly Farm as it is preserved today. *Courtesy of the author.*

influential on the eastern side of Watchemocket that the surrounding area came to be named "Kent Heights." Another long-standing farm was established on the lane off Pawtucket Avenue called Willett Avenue. The 1805 Federal-style house built on the property by sea captain Pearce Allin still stands and is on the National Register of Historic Places. When a farm was established there late in the century, it quickly became a popular farm stand for those returning from a day at the "resorts" that grew along the bay in the nineteenth century. Many family photographs were taken on the expansive lawn of the farm, whether posing with the friendly looking, cast-iron sculpture of a Dalmatian that became a favorite of children visiting the farm or beside the granite pillars that still grace the entrance before Whitcomb Farm.

The town of Warwick, in 1875, contained 302 farms, whose lands took up nearly 75 percent of the town's land. But most of these farms were small; only 5 percent of farms were more than two hundred acres. Dairy farming was a major part of the town's agriculture, and much of the farmland was left as meadow for grazing or growing hay and corn for the cattle. Poultry and produce were also an important part of the local agriculture. In that same year, the town was fourth in market garden production. The Arnold Farm, which had been established in 1642, remained an active local farm until 1872, when it was sold and many acres divided into lots for housing. Shortly after, Robert Perry purchased the house and substantial acreage to establish a dairy farm. The successful business was leased to the Leach family by 1914, and they continued the operation for several generations.

Nearby, Jeremiah Battey had established a stock farm in Pawtuxet, building a house and large barn in the latter half of the eighteenth century. By 1795, he had sold the farm to his brother William. The farm passed through several hands until Jason Reynolds purchased the property and ran a popular tavern named the "Golden Ball Inn" for twenty-seven years. By 1871, Cyrus Cole had purchased the farm and established a dairy and ice business. After his death in 1886, his son Frank continued the business. Being active as well in local affairs, Frank Cole became instrumental in improving the local roads and transportation, paving the way, so to speak, for his son to carry a retail dairy business into the next generation. An article outlining the history of Cole's Farm appeared in a local newspaper in 1932 and declared of the modern farm that "a recent inspection of the dairy revealed the fact that the most modern equipment is used in handling the milk and the sanitary rules are rigidly enforced."

The Foster Farm on Centerville Road was a small dairy enterprise. The 1850 census shows that John Foster, then age forty-five, ran the eight-acre farm with his wife and two children. Their livestock consisted of two oxen, a horse, five swine and six milking cows. That year, Foster grew seventy bushels of Indian corn and fifteen tons of hay for his cattle, producing six hundred pounds of butter.

Despite the post–World War II suburban building of houses that covered more and more of the old farmlands that had been divided into lots and sold by families, parts of the city remained rural well into the twentieth century. An article in the Providence *Evening Bulletin* in 1941 read:

> *Officially, Warwick is a city but it has not quite made up its mind whether to become a full-fledged city or remain the half country town and half city it is today…Warwick usually presents a picture of fully developed settlement on one side of the street and a dairy farm or market garden on the other. The dreams of many a city worker are interrupted in the early morning by the crowing of roosters and the lowing of cows.*

Even today, lifelong residents of Warwick, such as Henry A.L. Brown, can remember the large fields of produce and pasturelands that covered much of the land well into the twentieth century.

The fields of local farms extended from Warwick Avenue and West Shore Road, and Cowesett Avenue as well. The Hoxie Farm had expansive fields that eventually abutted the Theodore Francis Greene airport by the 1940s. The Hoxie Farm was famous for its melons—the variety grown is unknown as the locals simply called them "Hoxie melons." These were picked and packed before the heat of day and shipped on the Warwick Railroad to Providence and, from there, to New York, where they would be on the market by next morning.

One of the largest farms in northern Rhode Island was the Wionkhiege Valley Farm. Established by a member of the Latham family in 1873, the farm housed a massive barn for cattle and supplied milk and produce to the Woonsocket area well into the 1940s.

Another long-standing farm still in existence today is Wrights Dairy Farm in North Smithfield. Established in 1896, by the turn of the century, George Wright was delivering milk to local families in milk cans. By the 1930s, his son Ernest had taken over the business, modernizing the farm for pasteurization, and establishing a delivery route for milk and cream delivered in glass bottles. The barn was passed along to his son Edward in

Wionkhiege Valley Farm barn and outbuildings. *Courtesy of the author.*

the 1970s, who opened a retail farm store to sell the milk and eggs produced. In 1972, Claire Wright began selling homemade pies in the retail store and this led the farm to expand once again, opening a bakery in 1976.

As the business began to grow, additions were made to existing buildings, and in 2001, a 120-cow freestyle barn was built. Eight years later, a new state-of-the-art milking barn was raised with large windows that allowed visitors of the farm to view the milking process.

Among the early twentieth-century buildings still in existence on the farm is the original dairy barn, with two small belfry-like ventilators.

The majority of dairy farms still existing in the state today send their milk to processing plants across the border to neighboring Connecticut or Massachusetts for pasteurizing, bottling and distribution. In 2004, with a $125,000 loan from the state, a group of five farmers began the Rhode Island Dairy Farms Cooperative to increase consumer awareness and consumption of local milk and other dairy products.

While the milk is still processed at a plant in New Britain, Connecticut, the product is transported throughout the state under the banner of "Rhody Fresh," and the cooperative's campaign to raise consumer awareness of local farms has paid off. By 2007, the business had increased by 30 percent, and nearly a decade later, the cooperative has grown to nine farms and has seen a revenue of $3 million from the production of milk and other dairy products. In 2013, it was announced that the cooperative had been awarded a $150,000 grant from the U.S. Agriculture Department to upgrade its production facilities.[67]

Louis Escobar of Highland Farm in Portsmouth, one of the founders of Rhody Fresh, sums up the mission of the cooperative and the vision of a hopeful future for Rhode Island's dairy farmers:

> *With Rhody Fresh, you are not only buying milk, but also helping to preserve our farms, the open spaces, and the rural character in our communities throughout the state.*

In 2013, the 225-acre Bailey Brook Farm was chosen by the Rhode Island Green Pastures Committee as the Outstanding Dairy Farm of the year. The farm, operated by Rodney Bailey and son Paul, was cited for its "outstanding relationship with the community, use of good management practices, and commitment to ensuring a viable agricultural industry in the West Bay."

The farm sits on historical property, as part of the Fry Hamlet Historic District along Route 4 in East Greenwich. Presently owned by three siblings—Rodney and Gladys Bailey from the town, as well as sister Priscilla Crofts of North Stonington, Connecticut—the farm holds a herd of eighty Holsteins and Jerseys, which include forty milking cows. Farming has been a

Wright's Dairy Farm. *Courtesy of the author.*

long tradition in the family and has continued into the latest generation. Of the four children raised on the farm, son Paul followed literally in his father's footsteps, daughter Cynthia LaPrise established EMMA Acres Farm with her family—a recipient of the Green Pasture Award in 2010—and daughter Kathy Burroughs is currently president of the Rhode Island 4-H Club.

Rodney and his wife, Judy, have long been active in community affairs, being members of the Rocky Hill Grange for more than fifty years, with Judy serving on the East Greenwich Town Council for four years and eighteen years on the Rhode Island Agricultural Lands Preservation Commission.

Since 1980, the Bailey Brook Farm has been a member of the Agri-Mark Cooperative, a regional dairy cooperative that produces the Cabot brand of cheese and butter sold throughout New England. A member of the Rhode Island Dairy Farm Cooperative since 2009, part of the milk produced on the farm goes to Rhody Fresh products as well.

In its press release announcing the award, director Janet Coit of the Rhode Island Department of Environmental Management stated, "The success enjoyed by the Bailey family and other dairy farmers who produce local milk products is helping to protect and preserve hundreds of acres of farmland that will support continued agricultural endeavors for this and future generations to enjoy."

5

Urban Barns and Elegant Carriage Houses

With the prosperity brought to Providence and other urban centers by the China trade and later manufacturing in the nineteenth century, merchants, mill owners, military heroes and successful tradesmen alike began to build elegant houses along the crest of "College Hill" and on the main streets of what were once rural Rhode Island villages.

Along with these urban mansions, owners built ornate carriage houses and barns, often designed to mimic the features of the house. A walk along the East Side of Providence offers several examples of these well-preserved stables.

Beginning on Benefit Street, where the earliest of the community's elegant houses were constructed, we find the Colonel Joseph Nightingale House (1791), a three-and-a-half-story clapboard mansion with a gable-on-hip roof. The large stables, bay and carriage house were designed and constructed by architect Thomas Tefft in 1855[68] and derive in design more from the brick Italianite houses being constructed in Providence at the time. It is now called the Nightingale-Brown House in a nod to the additions added by the Brown family, including the carriage house and a wing devoted to an always expanding library that later became the John Carter Brown Library on the campus of Brown University.

A short walk to the terraced corner of John Street brings us to the Thomas F. Hoppin House (1853–55),[69] whose brownstone retaining wall once surmounted a cast-iron balustrade. The house is a grand Italianate manor with brownstone trim, and a short climb up John Street reveals the

Carriage house of the Nightingale-Brown house, now part of Brown University. *Courtesy of the author.*

carriage entrance and cobbled courtyard, where we see the two-story service wing, adorned with a wooden balustrade above three elegant arches, leading to the carriage house, whose lower windows and arched doorway continue the design from the eastern side of the house.

Continuing up John Street to Brook Street and turning left, a brisk five-block walk will bring you to the corner of Benevolent Street, and turning right, you'll find at 100 Benevolent a handsome brick carriage house with sculpted stone posts in front on the left. This was once the carriage house for one of the Goddard family's houses nearby. It holds a brick archway above black paneled doors and an elegant glass wind-eye above. Stroll up to the corner of Hope Street, and you'll see the rear of the massive carriage house and stables that were built for the Sprague-Ladd House, constructed by the Sprague family in 1850. Though the two-and-a-half-story, U-shaped, stone-trimmed brick house was modified in 1902 by G.F. Ladd, the stables were left unscathed.

Continue up Benevolent Street to Cooke Street, and the Benjamin B. Adams House (1871–72) will be on the corner. It is a two-and-a-half-story Italianite house—a style that had gained much favor in Providence during the affluent years after the Civil War—with an adjacent brick cross-gabled-

View of the service wing and carriage house of the Hoppin House, now part of Brown University. *Courtesy of the author.*

The expansive stable and carriage house of the Sprague-Ladd House, now part of Brown University as the Orvis Musical Center. *Courtesy of the author.*

roof stable and carriage house. Set back at the end of a formal courtyard off Benevolent Street, the carriage house is almost chapel-like at first appearance, with its steeple-like cupola atop the front gabled roof and brick balustrades at each end. Three bay doors face the courtyard, with an elegant arched window and hay door above the central wood- and glass-framed double door. A brick, sculpted motif mimics the house's grand façade, and a pair of black lanterns adorn each side of the central doorway.

Inside, though the bottom floor has been converted from stables to a clean, cement-floored garage for the house's tenants, the original floor and rafters are intact above. A wooden staircase leads to the second floor, where one finds that most of the original post-and-beam frame is intact, with few repairs. The true elegance of the hay door and windows is revealed inside, within a simple, arched wooden doorframe. The brick walls are whitewashed, and the floorboards are also, largely original. In the rear corner of the stable, a large pair of cast-iron hinges rest on the trapdoor lifted to drop hay to the stable below.

Continue along Cooke Street, cross and then turn left at George Street. At number 236, the carriage house behind the 1910 Federal home on the

The carriage house of the Benjamin B. Adams house. *Courtesy of the author.*

Interior of the Adams carriage house. *Courtesy of the author.*

site was designed by architect Richard Upjohn and built in 1860 with the Amos Smith House at 169 Hope Street (demolished in 1970).

Farther along, nearing Brook Street, we reach number 182 and the imposing rock-faced mansion built by prominent banker Henry Pearce in 1898. Preservationist and historian William McKenzie Woodward described the house as

> *a 2½ story, Richardsonian Romanesque house built of rock face granite ashlar with brownstone trim, this complexly massed structure with asymmetrical plan, has grouped, round-arched windows, a porte-cochere on the northwest corner, a copper and stained-glass conservatory on the west, and a conical-roof tower on the façade.*

An equally elaborate carriage house, complete with a conical-roofed tower sits across the road on Brook Street, disconnected from the house by the modern Computer Laboratory built in 1959 from a design by Phillip Johnson, a low-lying concrete and glass structure erected on the once expansive lawn. One can still follow the brownstone wall of the estate as it curves around Brook Street. Originally, a lane ran behind the lawn and connected with the carriage house across the street.

Continue west on George Street past Brown University's Sharpe Refactory, and turn left on Brown Street. Look to the right for a cobbled lane leading to the charming carriage house that once belonged to the Candace Allen House, down the hill at 12 Benevolent Street.

Walking down the hill, we reach Benefit Street once again. Turn left, and within two blocks, you are walking along the brownstone-enclosed, expansive lawn of the John Brown House. As you reach the intersection of Power Street, take a moment to cross Benefit and find the magnificent stable and carriage house that rests within a cobbled courtyard behind a cast-iron gate hung on sculpted granite pillars at the entrance.

Once the stables for the John Brown House on the hillside above, the original carriage house was renovated and enlarged by Marsden Perry in 1902, the owner of the Narragansett Electric Company and the city's trolley service. Perry had the stables designed in the Federal Revival style

The Pearce Carriage house, circa 1898. *Courtesy of the author.*

The Candace Allen Carriage House. *Courtesy of the author.*

The Marsden Perry Carriage House. *Courtesy of the author.*

Barn on Clarke lane in Providence. *Courtesy of the author.*

with a prominent cupola to complement the grand house he had just purchased. The stables housed Perry's horses, carriages and automobiles until his death in 1935.

In neighborhoods beyond the grand houses, more common stables and barns appeared. The simple, square, timber-framed barn on Clarke Lane was built for the service of the jeweler's large Italianite house on Keene Street.

The more common carriage houses, as mentioned in an earlier chapter, have long been converted to garages, but in many cases, the larger carriage houses still in existence have been converted to living spaces and studios—one is even an art gallery.

In the rural areas of Rhode Island, the increased use of stagecoach and carriage service precipitated the building of large carriage houses and barns, with farmers often constructing a building that could serve both purposes. The design of these, like earlier barns, was often based on simple, traditional plans like those printed in the *Carpenters Companion*, but in some cases, builders elaborated on the design. What historian Walter Nebiker

Carriage house and barn on Great Road. *Courtesy of the author.*

Stable and carriage house in Albion Village. *Courtesy of the author.*

termed "a classic example of a 19th century carriage house and barn" rests on Great Road in Lincoln, adjacent to the Ballard Farm.

Another example would be the carriage house built by Walter Allen behind the Stephen Brownell House in Union Village. Long in disuse, but still standing, the classic white, wide-doored carriage house with cupola is a ghostly reminder of earlier times as cars speed by on the old stagecoach route.

In the village of Albion, we find a lower, almost western-style structure designed to accomodate a good number of horses, with a large bay at the end for carriages entering and leaving. The building likely dates from the period when Albion Mills was operating and the village was at the peak of its population in the late 1800s.

With the advent of the railroad, these routes quieted quickly, and those stables, barns and carriage houses transformed into garages, blacksmith shops and tanneries. Some were used as small factories for beginning industries. For many years, while traveling the old routes of Rhode Island, you could see these great buildings' slow declines. Many were let go beyond repair and abandoned to eventually collapse in a storm. Today, those remaining are contained within the small packets of preserved historical areas along the old routes. Many are privately owned, and their preservation is solely in the hands of the owners. Fortunately, a good many of these owners in Rhode Island have made an effort to restore or simply maintain the barns and carriage houses on their property.

6

For Love of Farms and Farming

Carroll and Molly Harrington have lived on and operated a farm on Long Entry Road for over fifty years. The couple purchased the R.P. Colwell Farmhouse at the intersections of Tarklin, Colwell and Long Entry Roads, called for many years "Country Corners," in 1961 and spent the first five years getting the old farmhouse in shape for their family.

The fields were another matter. Having gone unplowed for years, they were overgrown with scrub, briars and poison ivy. The Harringtons turned to the U.S. Department of Agriculture's soil conservation program. When their application for assistance was approved, the department sent out massive surplus World War II Seabee bulldozers to clear the fields and, eventually, dig a pond, which Carroll says "was a tremendous asset," for the farm, wildlife and the family's recreation.

Beginning a farm was a learning experience for the couple. They read all they could and learned what they liked through time and effort. Molly had wanted to raise pigs, but when the couple had young children, they knew the two would not be a good mix. Carroll wanted to raise beef cows, so they started simple, with a pregnant cow and a calf. For thirty-seven years, they ran a stock farm for white-faced Hereford cattle, using modern artificial insemination to increase the herd a little more each year, without the cost and hazards of keeping a bull on the farm. At one time, they introduced a few Charolais cows to the herd, a French breed that Molly described as being extremely gentle.

Carroll and Molly Harrington. *Courtesy of the author.*

In the last years of the stock farm, they inherited a Scottish Highland cow, which weighed just under a ton and whose horns measured forty-two inches from tip to tip.

The barn on the Harringtons' property when they bought the house was a simple three-bay structure. The couple added extensions to house their cattle and have generally renovated the barn and other outbuildings on a continuing, as-needed basis. The barn, according to Carroll, has been "a work in process" for the fifty years the couple has lived on the farm. One renovation was to install 150 feet of black solar hosing on the roof that collects rainwater and feeds a water tank in the rafters of the barn.

The Harringtons also credit their neighbor, the late Alfred Hersey, a feed distributor for Agwam Inc., for teaching them a great deal as they started farming. "He really helped with everything," Molly recalled, "he would tell us what type of grass to plant and where." His advice helped to make the Harringtons' fields very productive and their cattle healthy.

The Harringtons raised four children on the farm, all of whom "loved growing up here," Molly says. One son now lives in Japan, where he recently began a public garden. They also hosted five exchange students on the farm through the "Academic Year in America" program over the years, an experience that Molly noted was especially gratifying for both the couple and the students. "They just flourished here," she told me, "there were things to learn and do that they'd never experienced before."

Today, the farm is operated as a hay farm for those local farmers who operate stock and dairy establishments in the area. It's an easier life, though Molly at eighty-one still climbs aboard the John Deere to clear a path for those that come to mow, bundle and haul the hay to their farms. Carroll continues to be an enthusiastic supporter of those who take a chance at reviving the old farms. One customer, the Blackbird Farm close by, maintains ninety head of Black Angus cattle and has become a prime provider of the favored beef to local restaurants.

Just down Long Entry Road from the Harringtons' farm is the old Steere Farm. Now called Long Entry Farm, the property has a lengthy history dating back to 1730, when William Steere established a dairy farm. The original barn held eighteen to twenty-four dairy cows, a tradition that was continued by the Mitchell family into the twentieth century.

After a succession of owners, Brown professor and pathologist Enold Henry Dahlquist purchased the farm in 1958. Dahlquist, according to his son Michael, was "a farmer at heart." He ran the farm to be self-sustaining for his family of seven children, with chickens, ducks, pigs and a pair of horses on the property. He put a new roof on the 1730-era barn and kept several acres productive over his lifetime.

The second-generation owner has now rejuvinated the farm by selling produce through one of the farmers' markets in the state, as well as participating in the community-supported agriculture program, allowing him to sell fifteen shares of crops each season. He also sells hay to neighboring farms and is more than willing, in this still tightknit community, to lend a hand when needed. On the afternoon that I visited with Michael and his family on the farm, he had spent part of the previous day helping a neighboring farmer harvest the hay crop when his equipment had broken down.

Michael would like to expand the farm even more, taking back more of the fields that have long lain overgrown and expanding his vegetable gardens, perhaps even establishing a stock business with cows and pigs. He has restored a second barn on the property, whose stone foundation seems to have been built with the original farm buildings but has been modified and restored several times with both hand-hewn timbers and mill-cut posts and boards in evidence within the structure.

The original barn is now in disrepair, leaning precariously against the woodshed addition from the 1970s with many of its vertical clapboards missing, leaving large gaps in the walls. Restoring the barn is a bit of a quandry for Michael, who repects the history of the farm and its buildings but faces the restrictions such restoration would mean on a modern farm. Were he to restore the barn to its eighteenth- and nineteenth-century structure, it would be little more than a decorative period piece on the property with little use to store the equipment and supplies now needed on the farm. As

Michael reflects, it's the main reason why so many old barns "have fallen by the wayside."

I told Michael of Eric Sloane's great observation in his *Weather Almanac* about such old barns:

> *The wavy and rickety condition of many...is due directly to frost upheavels. Barns, which usually lie directly on ground stone foundations, often have roofs that have changed lines with the moving contours of the earth. Some of the barns with roof-lines that look like ocean waves are as firm as when they were built. If you look at the ground below, you'll find that the earth has acquired the same lines.*

The young farmer nodded knowingly and mused aloud about the storms and weather the barn had been through and acknowledged the builders who had constructed such a barn.

Having grown up on the farm and now striving to maintain a small living from the land, Dahlquist is grateful for the opportunity to follow in the footsteps of those neighbors he got to know as a boy. He holds a tremendous respect for their way of life, as well as their perseverance and love of the land. To that end, Michael and his family sold twenty-two acres to Land

The barn restored by Michael Dahlquist on his farm. *Courtesy of the author.*

Trust so that, like the property of the former Seldom Seen Farm on Cooper Road, at least a part of the farmland will remain preserved.

Jim Dame of Johnston, Rhode Island, grows heartsick at the sight of abandoned farmland, but that is what he must face each morning as he looks out from his property on Brown Avenue and surveys the empty century-old farmhouse and the great barn and pathways of the farm that sweep down the hillside to the orchards of Dame Farm. All lay silent and vacant in the summer of 2013.

The farmland was originally purchased from the Steere family by Donald Dame in 1890. The Dames planted apple orchards and raised dairy cows on the farm, which continued for more than seventy years, until 1969, when the state exercised eminent domain and took over ownership of the land as part of the "Green Acres Project," a public parks initiative that eventually led to the development of the 744-acre Snake Den State Park. Although James Dame Sr. and his wife, Effie, were allowed to live in the old farmhouse, the land lay fallow for several years.

The Rhode Island Historical Farm Association (RIHFA) began renting part of the farmhouse to use as a museum and meeting place for group tours of the historic property. The Dames began subleasing the land from RIHFA in the early 1970s, and this allowed them to resume the running of Dame Farm. For the next two decades, the Dames ran a heifer replacement business in which they raised heifers and then bred them before selling the cows to local dairy farms. They grew silage corn for the tall, twin silos outside the large barn and sweet corn for the farmstand, maintained the apple orchards and laid out great pumpkin patches for picking in the fall.

By this time, three generations of the family were working on the farm. The lease of the land was always tenuous, however, and the uncertainty forced some in the family to leave the farm for more secure pastures. When RIHFA folded, the family had to renegotiate the lease and come up with new and improved ways to earn income on the farm.

Over the years, the increase of visitors to Dame Farm meant working more to meet their needs, and the heifer business was phased out from the farm. The cows were replaced in the barn by three large draft horses named Bill, Bob and Dave, who became favorites of visitors to the farm.

Kristen Dame, a certified teacher, created an educational guide for instructors K-12 entitled "Life on the Farm" and led educational tours. Open to the public every day of the week, the activities provided for local schoolchildren and adults revolved around the farm's seasonal calendar. Strawberries were picked in spring, while hay was harvested and vegetables

The Dame Farm today. *Courtesy of the author.*

culled from the gardens during summer. Apples were picked in the fall, and pumpkins were chosen for the glowing jack-o'-lanterns on Halloween. Horse-drawn wagon tours of the farm ran from July through December, when, if the weather complied, visitors were treated to a sleigh ride over the blanketed fields.

All of that changed in 2013, however, when the lease that the Dames had held with the state was raised exorbitantly and out of reach of the family, who had lived and worked on the farm through seven generations. No one else has taken the state's offer to lease the land, and the farm now lies barren with low-slung chains on posts that now block the dirt roads that enter the farm from Brown Avenue. The only access left is for the walkers from the state park, whose parking lot lies off Route 6 in Johnston.

The farmhouse, barns and outbuildings remain on the National Register of Historic Places, but only time and budget changes will bring the Dame Farm to life for visitors once again. In the meantime, the family's tradition of farming continues with Jim's grandson, whose substantial vegetable gardens provide produce that is sold on the premises and in farmers' markets throughout the state.

There is no doubt, however, that the means by which farmers earn their living are changing yet again. Some have managed to find the resources to hang on through the difficult times and are emerging with a stronger, more competitive farm in today's marketplace, but others have fallen by the wayside or seen opportunity come too late. Such is the fate of Waterman Hill Farm, settled on long-established farmlands of the Waterman family, one of whose desecendants, Caleb Waterman, served in the Coventry militia as assessor of taxes and justice of the peace. He is listed simply as a farmer in the 1850 town census. The present farm at the crest of Waterman Hill was established in the 1940s by a man named Battey, who built a massive three-cupola barn on the property. The farm was purchased in 1961 by Jared Griffith, and he and his sons ran the dairy business until 1990, when the herd was dispersed.

One of the sons who talked with me told me that they'd lived their whole lives on the farm and that he'd milked a cow before he even went to school. The two brothers had great success over the years at tractor pull competitions, as a row of ribbons strung on the rafter in a shed behind us attested. He told me with a slight grin that these were "just a drop in the bucket." The farm is used for growing hay, stored in the great barn now empty of cattle. A recent grant proposal that would have restored the roof of the barn was denied, though whether the building was deemed too late a structure to be historic or the propsal denied for some other reason is uncertain. Nevertheless, the barn, to my mind, certainly remains a unique example of a large cow barn in Rhode Island.

Just down the road, the Greene Field Horse Farm is thriving on its fifty acres of old farmlands, providing riding lessons for kids and adults, as well as indoor and outdoor riding rings for practice in equestrian competition. The farm hosts the Ocean State Hunt Club Horseshows as well and is home to National Res Champion Fantastic Colors.

Around the bend from the Waterman Hill Farm, on Hall Road, the old Vaughn family homestead is being resurrected as a maple sugar farm, the present owners having been given a government grant to plant the maple trees that will bring the hoped-for future of the farm to fruition.

Similar stories abound throughout the state as second- and third-generation farmers find new uses for the land.

7
Preserving the Historic Farm

Prescott Farm, Middletown

The Newport Restoration Foundation

Though it is a sweltering day, a volunteer from the University of Rhode Island's Master Gardener Program invites us to browse the herb garden at Prescott Farm. While the herb garden is not "colonial" in the sense that a separate garden would not have been planted—early farmers mixed herbs within their vegetable plots—it contains, as much as possible, a plot of the herbs raised in the era, and the master gardener explains the myriad uses of each plant, be it cooking, medicinal or making early soaps and dyes.

In the distance, beyond a small grove of apple trees, lies the small, gambrel-roofed colonial clapboard house presumed to have been built around 1730 by the lieutenant governor of the colony, Jonathan Nichols, who owned the property at the time. The farm went through a pair of owners before 1771, when the house and farm were purchased by Newport merchant and distiller Henry John Overing, who was heavily involved in the Atlantic trade and a prominent slave owner. Overing was also a Loyalist, and as the conflicts between himself and other Newporters near his sugar shop on Thames Street grew heated, he might have retreated to his Portsmouth farm. Sometime after the British occupation of the city, which began in December 1776, Overing's farm became the headquarters of British general Prescott.

The Prescott Farm, as preserved by the Newport Restoration Foundation. *Courtesy of the Foundation. Photo by the author.*

It was in this colonial farmhouse that the general was captured in a daring raid led by the young Captain William Barton of Warren, who was then encamped in Warwick, Rhode Island. Barton led five rowboats of men onto Narragansett Bay and rowed between Patience and Prudence Islands to avoid detection by the British ships, landing at the Bristol Ferry close by the house, where they captured the general and his aide-de-camp in their nightclothes and rowed them back to Warwick before daybreak.

This singular historical event so took hold on the property that since the mid-nineteenth century, it has been called the Prescott Farm. The event also may have saved the farm from developers. Recognizing the historical importance of the farm, the property was purchased at a crucial time by Doris Duke, the leading philanthropist for the presevation of historical Newport property. In 1970, the year after its purchase, she deeded the house to the Newport Restoration Foundation, which has run the farm as a historically preserved farm ever since.

Partnering with the Master Gardener's Program of the University of Rhode Island has ensured that the gardens that would have been grown on a self-sustaining colonial farm are replanted each year and maintained by volunteers. The Prescott Farm also offers gardening lessons and educational events during the year. The farm is surrounded by period houses, moved to their current location by the NRF to save them from demolition. These include the Hicks House (circa 1715), a simple salt-box cottage with a slanted roof that housed the family who ran the Bristol Ferry nearby; the Sweet-Anthony House (circa 1730), a larger slant-roofed structure that includes a few nineteenth-century additions to a simple farmerhouse; and the "guard-house," built as an addition to the Nichols-Overing House in 1840 and removed to restore the historical integrity of the original house. The simple, gambrel-roofed structure is now used as the visitors' entry center to the farm museum.

The Prescott Farm's most famous attraction is undoubtably the tall, smock-style, four-vaned windmill constructed by Robert Sherman for use with his distillery in Warren. The windmill was moved to Portsmouth and used in separate locations by two millers until the twentieth century, when it fell into disuse and disrepair. The NRF purchased the windmill in 1970 and moved it one last time to the farm. It was restored in 1971, with further work being done in the 1980s, as well as the installation of a new shaft in 1998. The windmill remains a fine example of the millers' industry on Aquidneck Island.

Watson Farm, Jamestown

Historic New England

I arrived early at the Watson Farm in Jamestown, and all was silent as I got out of the car, but for a few crows arguing in the wooded edge of a nearby field. A friendly border collie lying near the farmhouse spied me and bounded down the hillside to greet me. The farm contains a clapboard Federal-style house with an addition that dates the existing structure to 1796, while the barn across the wide dirt lane dates from 1805, with a later nineteenth-century addition. A cow barn on the property also dates from the nineteenth century, and a later 1950s-era barn is well hidden behind a copse of trees.

I found Don Minot, the manager of the farm, ready to start up the tractor to cut a field before our scheduled time, but he's an amiable man and invited me to sit down in the shade while we talked. Don and his wife, Heather, have managed the farm for thirty-four years on behalf of Historic

The Watson farmhouse. *Photo by the author. Courtesy of Historic New England.*

New England, which acquired the 265-acre property in 1979 with the stipulation that it remain "a working farm." The Mintos have been the sole managers since the property was preserved.

He calls the job "a great opportunity to stay on this beautiful farm," and it also means a committment to improving farm paractices, the food produced for consumers and the improvemnet of the land itself as farmers. The task was not easy at the start, for among other issues—one being that the house had not been lived in since the '38 hurricane—there was a feral herd of one hundred Angus cattle on the property. Fences had long fallen apart, stone walls crumbled into disrepair and the cows and bulls were everywhere.

Eventually, they got the herd under control and ran a calf and cow operation, but the cattle had become too inbred, and calves were often underweight. They integrated the herd with Simmental bulls to "beef up" the cows, and while the calves grew bigger on the larger frame acquired with these bulls, they were not well suited to being purely grass-fed cattle, which is what the Mintos sought to have on the farm. The idea of grass-fed beef was not a popular idea in the 1980s, but Don also knew that with 180 acres of grasssland on the farm, it was a perfect situation to bring the idea to fruition.

The couple began to look at heritage breeds: the Scottish Highland cattle; the Jersey dairy cows; and the red Devons, which were said to be the first to graze on the grasslands off Narragansett Bay. It seemed a natural choice, and today, the farm holds one hundred head of cattle, mostly the red Devon breed, with a few Black Angus mixed in. The farm also keeps a large flock of sheep as well as ducks and chickens. Their hard work produces high-quality beef and finished lamb for local markets.

As caretakers for Historic New England, the Mintos want visitors to experience the farm as it has always been run—as a self-sustaining farm through the centuries—and are cognizant of the connection that many have with neighboring farms.

"People have grown up remembering the farms where houses now stand," Don says, and as a consequence, the awareness of preserving the land, as well as the agriculture, has grown. To that end, Don and Heather want visitors not only to see a working farm but also to experience it in other ways. The Mintos' daughter has taught a class on medicinal herbs that proved to be popular, with many home gardeners getting tips about plants and simple, often forgotten remedies. Heather provides educational programs and invites neighboring schools and organizations to utilize the farm for educational classes that reconnect people with the land and, hopefully, come to know the importance of it and the farmers who have worked these acres for generations.

Don, like many who work a farm, sees the growth of farmers' markets as an opportunity for local agriculture to grow, as well as an opportunity for young people who, like himself more than thirty years ago, dreamed of running a farm. An intense student of farm management techniques over the years and the holder of a soil conservation degree from the University of Rhode Island, he hopes that what they have done at Watson Farm in rotational grazing and innovative methods of soil fertalization can be a role model for future small farmers.

He also has seen the benefits, firsthand, of working with an organization such as Historic New England and the impact of Thomas Carr Watson Jr.'s gift to the community. In the thirty-plus years since Watson ensured the continuity of the land and agriculture on the farm, the community has responded by preserving over 1,200 acres of land and an additional two hundred farms on Conanicut Island through organizations such as the Nature Conservancy, the Conanicut Island Land Trust and the Grange's Farm Viability Committee.

CASEY FARM, SAUNDERSTOWN

Historic New England

Just across the bay from the Watson Farm lies the historic Casey Farm, also preserved by Historic New England. Owned and managed by the organization since 1955, the farm has come to represent a myriad of resources, having long provided educational programs, summer camps, a Boy Scout encampment and what a farm provides best: locally grown produce to the community and visitors from throughout the state.

When I arrived with Dan Santos, the regional manager for Historic New England, the farm was teeming with children playing on the expansive lawn before the great 1750 house. Other groups were huddling to listen to a staff teacher tell them that the black-and-white feathered chickens scratching around the pens are Dominiques, the oldest breed of chicken in America; or watching the swallows swoop into the horse barn, where nests lie in the rafters above the stored hay in the loft and the stalls were converted to heated chicken bays, where more Dominiques and Rhode Island Reds are growing. One popular program at the farm is "Project Chick," where a farm educator visits a classroom with everything needed to incubate a clutch of eggs, instructs the classroom, sets up the equipment and leaves the clutch in the care of the class. Twenty-one days later, the eggs will hatch, and a few weeks later, the students bring the chicks to the farm, where they learn more about this endangered species and their lives on a free-range farm.

The large house that Daniel Coggeshall Jr. built for his family now hosts art and educational sessions for adults and children and also has a front room set aside as a museum of the farm. Inside, prominent portraits of the Casey family line the walls, from the staid, colonial portraits of Silas Casey and his wife, Abigail, to the formal portrait of Thomas Lincoln Casey and the less formal nineteenth-century photographs of the family on the farm. Perhaps most poignant of all are the prints from glass-plate negatives portraying daily life on the farm. These photographs were taken by Harry Weir Casey, the son of Thomas Lincoln Casey, the career military officer who managed the farm in the mid-nineteenth century and is credited with restoring the eighteenth-century house, as well as building the current carriage house and wood shed; the tobacco barn, later used as a corn-crib; and the pig-house-now used as a storage shed on the farm.

While Harry's photographs give a physical glimpse of the nineteenth-century working farm, his letters are full of escapades and complaints about the tenants on the property. Harry entered the scientific program at Yale College in 1870 and won several promising awards before tragically drowning three summers later off Narragansett Pier. Thomas Lincoln Casey put his reflections of the farm in the book *A Historical Sketch of the Casey Farm* (1881). In its pages, he makes clear that the work of rejuvinating a farm whose soil had become "exhausted in fertility," whose fields were stony and whose mansion house was "dilapitated" was accomplished not only to preserve a family homestead but above all else for his "love for the land." In addition to the improvements made to the buildings on the farm, Thomas Lincoln Casey also planted one hundred American elm trees on the farm, many which lasted until the hurricane of 1938, and improved the cranberry bog in the swamp meadows on the property.

The cow barn on the Casey Farm, circa 1850. The ell to the right held calves separate from the cows on the left. They were herded through a narrow four-foot-high door in and out of the main structure. *Courtesy of Historic New England.*

I met Ashley, the young manager of the farm who oversees the operations and seems clearly to be the eye in the center of the storm on days like these. She spent some time with Dan and I, explaining how the farm is currently run to remain a viable source of locally grown produce. As part of the community-supported agriculture (CSA) program, the farm offers customers a host of produce grown on the farm—lettuce, summer squash and tomatoes—as well as eggs from the 250 hens on the property. CSA members of Casey Farm may also get a discount on their produce if they volunteer at the farm and give "hands-on" help with various tasks.

Casey Farm hosts the Coastal Growers Farmers' Market every Saturday from May to October. The farm is also host through the summer for several of the herd of red Devon cattle from the Watson Farm that are let into the grazing fields across Route 1A in Saunderstown.

SMITH'S CASTLE, COCUMSCUSSOC, NORTH KINGSTON

Cocumscussoc Association

There is little left on Cocumscussoc that tells the long history of the place, but the house, built in 1678 after the original trading post was burned during King Philip's war, still stands. One can see from the site on the beautiful Mill Cove that it must have been an ideal location for Smith and Roger Williams to have established their place of business. The peaceful, farmlike existence of Smith's Castle today masks the years of land disputes with Connecticut and Massachusetts, as well as that bloody period on Richard Smith Jr.'s farm when he invited militia from Massachusetts and Connecticut to camp out on the estate he had inherited from his father. The resultant massacre of the Narragansett, as well as elders, women and children of the Wampanoag at the Great Swamp, led the Narragansett to burn to the ground the elder Smith's house, a place where they had long traded, in retaliation. Forty colonial soldiers killed in the fighting were buried near the post—supposedly the site of a "grave apple tree." Today, a boulder and plaque mark the site.

Within two years, Smith Jr. had built the present farmhouse, which originally had an impressive two-story gabled porch at the entrance. A large, central stone chimney served the two front rooms with enormous seventeenth-century timber-mantled fireplaces. A lean-to kitchen, now modified as two small additions, finishes the house in back.

In the aftermath of the conflict, the farm came into the hands of the Updike family, the first of whom were Captain Lodowock Updike and his wife, Abigail, direct descendants of Richard Smith. This esteemed family transformed the farm into one of the largest plantations in the state, consisting of three thousand acres divided into five working farms, managed mostly by tenant farmers. Indentured servants, slaves—both black and Native American—and seasonal hired hands made up the workforce for much of the labor on the stock and dairy farms.

The Updikes, along with neighboring planters, also breeded Narragansett Pacers, as well as growing some agricultural crops. Abigail Updike would outlive the rest of her family and remain on the farm until her death in 1824.

The property was then inherited by Benjamin Congdon's heirs, and the Congdon family—mainly the women of the family—ran the farm for forty years. The young men, who one would expect to work the farm, died in their twenties and thirties. Cousins and uncles helped out, and parcels of land were sold over the years to gain income, but Homestead Farm, as it was called, was managed for much of its existence by Eliza Congdon with help from Sarah Allen and Martha Spink, both relatives who would eventually inherit the farm.

After a succesion of owners, the farm would again be resurrected when it was purchased in 1879 by Anna Babbit. Mrs. Babbit made extensive modern renovations to the exterior of the house, masking with Victorian style the simpler colonial lines of the old farmhouse. She constructed a large dairy barn, poultry house and outbuildings on the property. The farm was inherited by her daughter Alice Fox, the wife of attorney Austin H. Fox, and the couple ran a successful dairy until his death in 1937.

The great herd of purebred Ayrshires were disbanded, and the agricultural life of the farm came to a close. The farm soon came to be neglected and was threatened with demolition. In 1948, the Cocumscussoc Association was formed to preserve what remained of the property and assure its use for public education. The house had undergone some restoration when Fox hired Norman Isham to perform minor renovations, one being the uncovering of a fireplace that had been covered for nearly a century. Smith's Castle was most recently restored by the architectural firm of Clifford M. Renshaw, for which it won the 1998 National Preservation Award from the National Trust for Historic Preservation.

Smith's Castle contains artifacts from the Updike and Congdon families, as well as an impressive library and extensive gardens containing plants and flowers that would have been held in eighteenth- and nineteenth-century plots.

As one of the few "living museums" in Rhode Island, the house is host to a wealth of educational programs provided by the Cocumscussoc Association throughout much of the year. When I visited the farm, an exhibit and program on pirates of the Red Sea trade was on display. These pirates frequented Rhode Island waters while waiting to weigh anchor for their next adventure, as the colony was well known as a refuge for wayward mariners. A seasoned actor played the pirate William Kidd, and a group of rogueish-looking, but merry men and women sang sea shanties there by Cocumscussoc's shore.

Other recent programs have included reenactments of the daily life at the seventeenth-century trading post as part of a series called "Bridge to the Past." The association has also hosted presentations about the fascinating lives of the occupants of the farm, including a review of the career of General Daniel Updike and the role of the Smith and Updike families in the institution of slavery in Rhode Island, as well as the lives of the slaves and indentured servants who lived and worked at Cocumscussoc.

Smith's Castle also hosts yearly Strawberry Festivals, Harvest Festivals and holiday celebrations. Tours are provided from May to December. The house, grounds and location on the bay alone make the visit worthwhile.

METCALF-FRANKLIN FARM, CUMBERLAND

Metcalf-Franklin Farm Preservation Association, Friends of the Franklin Farm

In the town of Cumberland, preservation-minded citizens have found a model way to save a historic property and use the land to benefit communities throughout the state. Since 2005, on the site of the former Metcalf-Franklin Farm on Abbot Run Valley Road, groups of volunteers have cultivated, seeded and tended the Franklin Farm Community Garden. In that time, the garden has produced 130,000 pounds of vegetables for community services such as the Rhode Island Food Bank, the Pawtucket Soup Kitchen,

Blackstone Valley Emergency Food Center, the Salvation Army Food Pantry and many other church and local soup kitchens and shelters throughout the state. Hundreds of volunteers have shown up for "tending nights," June through September over the years, and the efforts of the group were recently documented by Capital Television for the Rhode Island General Assembly.[70]

Long used as a dairy, the Metcalf-Franklin Farm sits on sixty-five acres of meadow and woodlands, with a preserved English-style barn from 1800 and colonial farmhouse built in 1854 on the property. Tending nights in 2013 were held Monday and Thursday at 5:30 p.m. Anyone may lend a hand by showing up and finding a team leader among the friendly volunteers who will assign a plot that needs picking. I can think of no better way to spend a summer evening than reconnecting with the simple rewards of farming by volunteering on this beautiful old farm and sharing that reward with others.

Chase Farm, Lincoln

Gateway Park, John H. Chafee Blackstone River Heritage Corridor

Visitors to the Chase Farm and Gateway Park, at the entrance to the federally protected John H. Chafee Blackstone River Heritage Corridor on Great Road in Lincoln, Rhode Island, have the opportunity to get a unique glimpse into the evolution of a farming community from the seventeenth to the late twentieth century.

The large tract of land whose remnants are preserved was once the homestead of Thomas Arnold and his purchase of 1661. The oldest existing house is that of his son Eleazer Arnold, whose massive stone-ender sits at the intersection of the Great North Road from Providence. Built in 1693, the farmhouse sat on sixty-five acres of the original two hundred plus acres that were allotted to family members by the end of the seventeenth century. Arnold increased his landholdings to seventy-five acres and with his wife raised seven children on the farm. He owned five dairy cows to produce milk, cheese and butter, as well as two oxen for heavy labor in the fields, nineteen sheep, a small herd of swine and a horse. Eleazer Arnold also grew a cash crop of tobacco and maintained an apple orchard, from which he processed vinegar and cider for visitors to his home.

The Chace Farmhouse, circa 1790. *Courtesy of the author.*

The Arnold House had a long and favorable reputation as a tavern during the eighteenth century, with its large "great room" and fireplace.

In the nineteenth century, the house became a much-photographed symbol of early Rhode Island, which no doubt helped the cause of its preservation. In 1919, the Arnold House was donated by Preserved Whipple Arnold to the Society for the Preservation of New England Antiquities (now Historic New England). The society hired noted architect Norman Isham to stabalize the house and assess its condition the following year.

During this same period, the Croade Tavern (1700) was moved from Pawtucket and placed behind the house. Here, Isham's team worked out their plan to shore up the old structure. Thirty years later, a more extensive renovation was undertaken in an effort to remove modern additions and restore the house to its seventeenth-century appearance. This included the construction of period-like diamond casement windows that had originally adorned the house, as well as replacing interior paneling and adding a plain, wooden door. Eleazer Arnold was one of the wealthiest farmers in the area and the Arnold family so extensive there that it was known as "Arnoldia" for some time. A few of the Arnold houses remain. The expansive James Arnold House rests on the bank adjacent to Lincoln Woods, and the handsome, two-story, gabled, Federal-period Israel Arnold House circa 1740 was restored by owner Donald Hysko, who enlisted the help of Rhode Island preservationist Antoinette Downing. A two-year restoration project saved the house from irreversable irrepair, and it has been well preserved since that time.

Chase Farm was established in 1867, after Benjamin Chase had purchased land from the Arnold family to start a dairy business. The present, simple Greek Revival farmhouse was built in 1890 and suitable for the large family that worked the farm. As the dairy business was the family's livelihood, greater detail and expense went to the construction of a large barn complex, a stone dairy and other outbuildings, some of which remain today. A large fire ripped through the barn complex in 1925, so the large barns visitors see today were built after that period and come from the merger with the Butterfly Farm nearby. The Chase-Butterfly Farm operated until 1965, providing milk and dairy products for the communities of Lincoln, Albion, Manville and Greater Providence. The house sits adjacent to eighty preserved acres, much of which is rolling fields interspersed with corridors of woodland growth.

Adjacent to the Chase Farm park is the Moffett Mill, constructed in 1812 along the Moshassuck River, which weaves under the Great Road along its path to Providence. This small machine shop built by George Olney is believed to be the earliest in Rhode Island and is a rare example from the beginning stages of industry in Blackstone Valley. Initially, the mill produced and repaired tools for local farmers, but it expanded after 1850, when it was purchased by Arnold Moffet, who constructed the stone dam that visitors can still see today. Moffett increased production at the mill, as accounts show invoices for wagons, carriages and sleighs constructed, as well as parts for ships, machines for textiles, furniture and even coffins. At one time, the second floor housed a braiding factory for making laces for shoes and corsets.[71]

The Hannaway Blacksmith shop nearby is a former carriage house converted to a shop by William Hannaway, who had begun his blacksmith business in an addition to the Moffett Mill. In 1901, he purchased the carriage house and built a home on the property for him and his wife. Hannaway was by all accounts a strapping, amiable and conversational man whose shop was visited often by neighborhood children who stopped to see him at work, setting wheels for carts and wagons or shoeing horses. With the advent of the automobile, the business declined, and after

1920, Hannaway mostly used the shop for shoeing horses for the nearby Butterfly Farm riding academy. Today, the blacksmith shop is open for demonstrations of the old craft to visitors on weekends, in which local smiths create hinges, nails and other household items forged during the mill's operation. The wheelwright shop also remains to demonstrate the repair of wagon and carriage wheels. Limited smithing classes are even held on the property.

The impressive stone mansion called "Hearthside" that sits back from Great Road was built just for that purpose—to impress a young woman from a prominent Providence family who its builder, Stephen Hopkins Smith, had long hoped to marry. Local lore tells the story of how Smith, a humble Quaker, fell in with the social circles in the city and there met the girl whom he began to court. Smith was not from a wealthy family, as his hoped-for bride-to-be was, but when he won a lottery of substantial size, he began to construct the house he hoped would match the luxurious lifestyle his girl was accustomed to living.

He kept his winnings and his plan a secret. The house was completed in 1814, and according to the story, Smith soon brought his would-be bride from Providence to view the house, who, on seeing the beautiful stone mansion, was said to exclaim, "What a beautiful house, but who would ever want to live way out in the wilderness?"

Smith was heartbroken, and the relationship soon fell by the wayside. He moved into the house with his brother George's family but soon tired of the arrangement and moved to another house down the road. He placed his energy and wealth into other projects, building a stone mill across from Hearthside and investing with Moses Brown in the building of the Blackstone Canal. His real legacy, however, would be in the great gardens and landscaped areas he created on lands he owned called Quinsnicket, a Native American word meaning "large stone houses," and a place used by generations of Indians as an encampment.

Smith filled Quinsnicket with stone walkways, waterfalls and ponds. He planted exotic trees, shrubs and bushes obtained in the China trade on the 458-acre tract. He also planted English ivy at Hearthside and two tulip trees that graced the walkway to the front of the house. The gardens around Hearthside were equally exotic, and in the early twentieth century, they became a favorite haunt of the poet and science fiction writer H.P. Lovecraft.

Stephen Hopkins Smith died in 1857, and more than fifty years after his death, the Metropolitan Park Commission met at Hearthside to finalize the purchase of the great house and the adjoing lands of Quinsnicket, which formally became Lincoln Woods State Park in 1909.

For the visitor, a drive up the length of Great Road has its own memorable rewards. This sanctioned "scenic highway" contains a wealth of historical architecture within its mile from Smithfield Avenue to the old Washington Highway (116). Driving up the Great Road from the Gateway Park, we take a right and pass the great barn and silos of the Chase Farm. A short distance up the road on the right, a Federal house comes into view. This is the Inspersion Jenckes House, circa 1762. Recently restored, this house remains a fine example of the period, as elegant in its wooden clapboard sheathing as Hearthside is in stone.

As Great Road winds uphill and then levels off, the names of the new roads like Bear Creek Court and Great Meadows Lane, are testaments to the farms that made up much of this land. On the right at the intersection of Simon Sayles Avenue lies an elegant Greek Revival house with a Doric portico that was originally the Smithfield Limerock Bank (1834). It soon ceased operations, and the house became a private residence. The Great Road curves left, and as it straightens, the Nate Mowry Inn (1810) and restored barn on the Ballard Farm come into view on the left along with a few small houses and buildings across the road that belong to the same period. A short distance farther, on the corner of Anna Sayles Road, lies the Mount Moriah Lodge. Originally a one-room schoolhouse, the building was doubled in size and encased with brick in the nineteenth century, and it remains, as it has since its beginning, set back from the road with a large dirt lot in front.

On the left side of Great Road stands the wonderfully preserved mid-nineteenth century carriage house and barn mentioned earlier in this book. Bearing right to follow the road to Route 116, one finds the old farmlands being divided and built on again, the stone-lined meadows plotted and populated by oversized, inelegant "McMansions." Just beyond this valley, however, lies the Valentine Whitman House (1696), a practical dwelling constructed at the closing of the seventeenth century and equally as impressive as the Arnold House at the beginning of the Great Road.

COGGESHALL FARM

Bristol, Rhode Island

A gaggle of geese greeted me as I strolled along the lane to Coggeshall Farm, the eighteenth-century preserved farm adjacent to the later and larger estate that is now Colt State Park. "Greeted" is a kind word to use, as the geese were true to their ways, quite aggressive with "intruders," forcing them clear across the other side of the road, hurriedly skirting the edge of the stone wall as they passed. "We call them the old ladies," mentioned one passerby. Nevertheless, they parted when I walked uphill on the dirt drive onto the farm. At the Coggeshall Farm Museum, the visitor witnesses an example of the post-Revolutionary agrarian life that was common in Bristol. As a self-sustaining farm, one sees the everyday tasks, the husbandry and gardening as they were done on the farm.

The house on the present farm dates from 1799, and the simple three-bay barn and outbuildings are from the same, or a slightly later, period. Interpreters in period clothing perform the ordinary tasks while visitors roam the forty-eight acres of the farm. Educational programs are held on a year-round basis, and true to the museum's mission, they accurately reflect the seasonal activities of the period.

Each Saturday during the summer months of July and August, a "Wake-Up in the Barnyard" program introduces children to the farm animals and chores that children on the farm would face each morning. A hunt for fresh eggs, milking and brushing the cows and turning out the turkeys were all part of a daily routine.

Also during the summer, a "Kids in the Kitchen" program introduces visitors to recipes from colonial times, and novices can cook "johnnycakes" on the hearth or a recipe from seasonal produce as it was cooked two hundred years before.

Fall brings the Annual Harvest Festival, which features the food and crafts from local vendors that they might have made and sold in eighteenth-century celebrations, as well as live entertainment on period instruments and hay bale tossing, saw bucking and spirited seed-spitting competitions.

Hearth cooking workshops are also held. Using recipes from the 1796 edition of Amelia Simmons's *American Cookery*, participants prepare several recipes on the hearth from meat and produce raised on the farm and set the dining table as it would have been set for a family in eighteenth-century New England.

Winter brings chores inside and a brief onset of holiday spirit with the selling of hand-cut Fraser pines during the Christmas season. Even with the advent of the holiday, life on the farm continued, and visitors get a brief glimpse of the hard life it once was during the winter: chopping and stacking wood for the fires to cook and stay warm, keeping the well clear of ice to draw fresh water and the keeping of preserves and meats to last the season, which were all necessary to carry the farm through the cycle to spring.

It is in preserved farms and places such as these that people can reconnect with a lifestyle that was so much a part of our past and has stood as a model of hard work and diligence for generations of Rhode Islanders. Throughout our history, even as we went to sea or traveled by other means to some far-off place, we have looked back at the farm.

As industry's promise became a life of drudgery for many, Rhode Islanders returned to the farm. When duty to country called during two world wars and farms were left to mothers, wives and widows to manage, those men who did return often returned to the farm. In our own hurried, high-tech times, many young people in the state have begun to see the simpler, hardworking life of farming as a welcome alternative to the endless rote of consumerism, meetings and media in our new 24-7 society.

In that spirit I suggest that the visitor to Rhode Island's historic farms enjoy the history and the charm of these wonderfully preserved places, and be sure to stop at a roadside stand on your way back to the city. Once home, frequent those farmers' markets selling local, organic produce, beef and dairy and purchase locally produced milk and dairy products in the supermarkets. You'll be supporting farmers and their families who continue a time-honored tradition in our state.

Notes

Chapter 1

1. The formal lots for Providence were not drawn up by Roger Williams until two years after the settlement's founding, at the insistence of the land owners. See Lafantasie, *Correspondence*, editor's notes.
2. Lafantasie, *Correspondence*, 211.
3. Providence Record Commissioners, *Early Records of the Town of Providence*, 1:3.
4. Chapin, *Documentary*, 17.
5. Ibid., 80.
6. Bridenbaugh, *Fat Mutton*, 13.
7. Ibid., 14.
8. Ibid., 45.
9. Ibid., 39.
10. Sloane, *Our Vanishing Landscape*, 19.
11. Ibid., *American Barns*, 60.
12. Ibid., *Age of Barns*, 22.
13. Ibid., *American Barns*, 60.
14. Ibid., *Weather*, 112.
15. *Historic and Architectural Resources of Burrillville*, Walter Nebiker, "Preliminary Survey Report—Town of Burrillville," 72.
16. *Historic and Architectural Resources of Burriville*, Preliminary Report, RIHPC, 1982.
17. Sloane, *American Barns*.

18. Mary A. Harris, "Life In Old Narragansett," *Rhode Island Historical Society* 19, no. 2 (April 1921).
19. Sloane, *Our Vanishing Landscape*, 40.
20. Ibid., *American Barns*, 44.
21. Ibid., 46.
22. Hazard, *College Tom*, 29.
23. McBurney, *History*, 8.
24. Miller, *Narragansett Planters*, 41, footnote Carrier, from *Beginnings of Agriculture in America*.
25. McPartland, *History*, 30
26. Architectural preservationist Norman Isham dated the house at around 1715 in the 1890s. This is disputed in the Rhode Island Historical Preservation Report nearly a century later, which states that the original house was built "well before" Isham's date.
27. Preliminary Survey Report of Historical Resources in the town of Warwick, RIHPS 1981.
28. I am indebted to Elsie Taylor, the present owner of the house, for the tour she gave me. Mrs. Taylor and her husband have restored the house as close to its original condition as possible over their thirty-plus years of owning the house and have always welcomed visitors, including members of the Greene family, who occasionally seek out the house.
29. *Historic and architectural Resources of Warwick*, Preliminary Survey Report, RIHPS, 1981.
30. *Historic and Architectural Resources of Scituate*, RIHPS, 45.
31. *Historic and Architectural Resources of Coventry*, Preliminary Survey Report, RIHPS, 1978.
32. *Rhode Island Historical Society* 5, 127.
33. Nebiker, *History of North Smithfield*, 17.
34. Ibid., 18.
35. Some readers will no doubt be familiar with East Providence's Hunts Mill, established before King Philip's War and then rebuilt and expanded during the nineteenth century. Its omission from this chapter results only from this area being part of Massachusetts until the latter half of the nineteenth century.
36. I am indebted to Christmas Moore of Brown University for her detailed description of the barn. It is now in private hands but is visible from the entrance road to Pulaski State Park.
37. Cady, *Early Development*.
38. Miner, *Angell's*, 31–2.

Chapter 2

39. Hazard, *College Tom*, 71.
40. As printed in Miller, *Narragansett Planters*, 26–7.
41. Ibid., 31.
42. Philips, *Horse Raising*, 916–7.
43. Ibid., 922.
44. Hazard, *Recollection*, 65
45. Mitchell and Young, *American Husbandry*, 80.
46. Browne, *Letter Book*, 19
47. Hazard, *Recollection*, 65
48. See *Time* magazine from September 9, 1937 and "A Racing Life" by Allen Woodville (1969) for more on Seabiscuit at Narragansett Park.
49. Architect Peter Harrison's farm in Newport lost seventy fruit trees, and Hessian troops dismantled stone walls to be used as ballasts for British ships. See Bridenbaugh, *Peter Harrison*, 70.
50. Davis, *Gilded*, 12–13.
51. The Slocum family plot lies just to the left of the entrance off Glen Farm road. It is now a designated Rhode Island Historical Cemetery.

Chapter 3

52. From the Journal of William Rogers, RIHS, as printed in Geake, *Historic Taverns*, 34.
53. McPartland, *History*, 274.
54. Ibid.
55. I am indebted to Patrick Verdier for the photographs and recollections of the boardinghouse, one of which was written by his aunt.
56. Brown, *Painted Rooms*, 147. Ms. Brown's books are highly informative of the great old houses left in Rhode Island.
57. Woodward and Sanderson, *Providence*, 142.

Chapter 4

58. From Christopher Martin, "Rhodeside Dairies," Edible Rhody.
59. Hazard, *Recollections*, 13.
60. Ibid., 73

61. Hazard, *College Tom*, 90.
62. Olbert, *Kitchens*, 93–5.
63. Miller, *Narragansett Planters*, 36.
64. Hazard, *Recollections*, 16.
65. Rhode Island Department of Environmental Management, "RI's Farming Family Tree."
66. Wood, "History," 9.
67. *Providence Journal*, May 6, 2013.

CHAPTER 5

68. The descriptions of houses, cited architects and builders' biographies in this chapter are dependent on Woodward and Sanderson, *Providence* as well as Woodward's later *PPL/AlAri Guide to Providence Architecture*. The descriptions of the carriage houses are my own attempt to render as detailed a description as I am able.
69. This was originally the site of a Federal-style mansion constructed by John Innis Clarke. The house was owned by William Jenkes when it was destroyed by fire in 1840. Jencke's daughter Anna married Hoppin, scion of a family who became wealthy in the East India trade. Hoppin was a painter and sculptor and spent several years in Italy before getting married, which no doubt influenced his and architect Alpheus C. Morse's vision of the completed house.

CHAPTER 7

70. See Joseph Fitzgerald, "Greener Pasture," *Woonsocket Call*, August 15, 2013.
71. See "Hearthside History" from Friends of Hearthside website.

Selected Bibliography

Bridenbaugh, Carl. *Fat Mutton and Liberty of Conscience.* Providence, RI: Brown University Press, 1974.

———. *Peter Harrison; First American Architect.* Chapel Hill: University of North Carolina Press, 1949.

Brown, Ann Eckert. *Painted Rooms of Rhode Island: Colonial and Federal.* Warwick, RI: Spring Green, 2013.

Browne, James. *The Letter Book of James Browne of Providence Merchant 1735–1738.* Providence, RI: n.p., 1929.

Cady, John Hutchins. *The Civic and Architectural Development of Providence 1636–1950.* Providence, RI: Book Shop, 1957.

Chapin, Howard. *Documentary History of Rhode Island.* Providence: Rhode Island Historical Society, 1938.

City of Providence Record Commissioners. *The Early Records of the Town of Providence.* Providence, RI: City of Providence, 1892.

Davis, Deborah. *Gilded: How Newport Became America's Richest Resort.* Hoboken, NJ: Wiley, 2009.

Elia, R., A. Strauss and N. Seasholes. *Architectural Review and Phase I Archeological Survey of the Reconstruction of Post Road in North Kingston, Rhode Island.* Office of Public Archeology, Boston University, 1989.

Geake, Robert A. *Historic Taverns of Rhode Island.* Charleston, SC: The History Press, 2012.

Hazard, Caroline. *College Tom: A Study of Life in Narragansett in the 18th Century.* Boston and New York: Houghton Mifflin Co., 1893.

Hazard, Thomas R. *Recollection of Olden Times.* Newport, RI: Sanborn, 1878.
Knight, Sarah Kemble. *The Jounal of Madam Knight*. New York: American Book Company, 1938.
Lafantasie, Glenn. *The Correspondence of Roger Williams*. Vol. 1. Providence: Rhode Island Historical Society, 1979.
Latimer, Sallie W. *Narragansett by-the-Sea*. Dover, NH: Arcadia, 1997.
Lisle, Janet. *The History of Little Compton: A Home By the Sea 1820–1950*. Little Compton, RI: Little Compton Historical Society, 2012.
McBurney, Christian. *A History of North Kingston, Rhode Island 1700–1900*. Kingston, RI: Pettasquamscutt Historical Society, 2004.
McPartland, Martha. *The History of East Greenwich 1677–1960* East Greenwich, RI: Free Library Association, 1960.
Miller, William Davis. *The Narragansett Planters*. N.p.: American Antiquities Association, 1934.
Miner, George Leland. *Angell's Lane.* Providence, RI: Ackerman-Standard Press, 1948.
Mitchell, John, and Arthur Young. *American Husbandry Containing an Account of the Soil, Climate, Production, and Agriculture of the British Colonies in North America*… 2 vols. London, UK: J.Bew, 1775.
Nebiker, Walter. *The History of North Smithfield*. North Smithfield, RI: Bicentennial Commission, 1976.
Olbert, Michael. *Kitchens, Smokehouses, and Privies: Outbuildings and the Architecture of Daily Life in the Mid-Atlantic*. Ithica, NY: Cornell University Press, 2009
Philips, Deane. *Horse Raising in Colonial New England.* Ithica, NY: Cornell University Press, 1922.
Sloane, Eric. *An Age of Barns*. New York: Funk and Wagnells, 1967.
———. *American Barns and Covered Bridges*. Mineola, NY: Dover Publications, 2002.
———. *Our Vanishing Landscape*. Mineola, NY: Dover Publications, 2004.
———. *Weather Almanac*. Mineola, NY: Dover Publications, 2005.
Wood, Squire G. *A History of Greene and Its Vicinity 1845–1929*. Privately printed, 1936.
Woodward, William. *PPL/AIAri Guide to Providence Architecture*. Providence, RI: n.p., 2003.
Woodward, William McKenzie, and Edward F. Sanderson. *Providence: A City-Wide Survey of Historic Resources*. Providence: RIHPC, 1986.

Rhode Island Historic Preservation Commission Reports

Historic and Architectural Resources of Burrillville, RI. A Preliminary Report. Providence, RI, 1982.
Historic and Architectural Resources of Charlestown, RI. Providence, RI, 1981.
Historic and Architectural Resources of Coventry, RI. Providence, RI, 1978.
Historical and Architectural Resources of Little Compton, RI. Providence, RI, 1990.
Historic and Architectural Resources of North Smithfield, RI. Providence, RI, n.d.
Historic and Architectural Resources of Scituate, RI. Providence, RI, 1980.
Historic and Architectural Resources of Smithfield, RI. Providence, RI, 1992.
Historic and architectural Resources of Warwick, RI. Providence, RI, 1981.
State of Rhode Island and Providence Plantations Preliminary Survey Report: Richmond. Providence, RI, 1977.

Periodicals

Rhode Island Historical Society 19, no. 2 (April 1929).

Newspapers

Providence Journal. "Bailey's Farm Wins Outstanding Dairy Farm Award." May 6, 2013.
Woonsocket Call. "Greener Pasture." August 15, 2013.

Other sources

Edible Rhody website. www.ediblerhody.com.
Friends of Hearthside website. www.hearthsidehouse.org.
Rhody Fresh website. www.rhodyfresh.com.
Rhode Island Department of Environmental Management. "RI's Farming Family Tree." http://www.dem.ri.gov/programs/bnatres/agricult/pdf/oldfarms.pdf.

INDEX

J

K

L

M

N

O

P

Q

R

S

About the Author

Photographed at Chase Farm by Lauren Issa Paul.

Robert A. Geake has been writing about Rhode Island history and folklore for over thirty years. His other books include *A Toll, A Tavern, and A Farm*; *Historic Taverns of Rhode Island*; and *The New England Mariner Tradition: Old Salts, Superstitions, Shanties and Shipwrecks* published by The History Press. His blog on Rhode Island history can be found at http://www.rifootprints.com. Mr. Geake is a member of the Rhode Island Historical Society and the Warwick Historical Society and is an associate of the John Carter Brown Library at Brown University.